Transport Perspectives for Porous Medium Applications

Edited by Huijin Xu, Chen Yang and Liwei Zhang

Published in London, United Kingdom

Transport Perspectives for Porous Medium Applications
http://dx.doi.org/10.5772/intechopen.111065
Edited by Huijin Xu, Chen Yang and Liwei Zhang

Contributors
Akira Nakayama, Anahita Asadi, Chen Yang, Cunliang Liu, Guojun Yu, Hanxing Zhu, Haokai Xu, Hesam Bazargan Harandi, Huijin Xu, Jianwei Guo, Liwei Zhang, Manguang Gan, Quan Xue, Shaicheng Shen, Ting Qiu, Xiaochun Li, Xiaohui Bai, Yixiong Lin, Yulin Wang, Yun Sun, Zhiming Fang, Zhongmin Wan

First published in London, United Kingdom, 2024 by IntechOpen
IntechOpen is the global imprint of INTECHOPEN LIMITED, registered in England and Wales, registration number: 11086078, 167-169 Great Portland Street, London, W1W 5PF, United Kingdom

British Library Cataloguing-in-Publication Data
A catalogue record for this book is available from the British Library

Additional hard and PDF copies can be obtained from orders@intechopen.com

Transport Perspectives for Porous Medium Applications
Edited by Huijin Xu, Chen Yang and Liwei Zhang
p. cm.
Print ISBN 978-0-85466-368-2
Online ISBN 978-0-85466-367-5
eBook (PDF) ISBN 978-0-85466-369-9

For EU product safety concerns:
IN TECH d.o.o., Prolaz Marije Krucifikse Kozulić 3, 51000 Rijeka, Croatia,
info@intechopen.com or visit our website at intechopen.com.

Meet the editors

Huijin Xu is a professor in the Department of Thermal Engineering, Shanghai Maritime University, China. He received his bachelor's degree and Ph.D. from Xi'an Jiaotong University, China. He worked in the Department of Mechanical Engineering, University of California, Riverside, USA, from 2016 to 2017. He worked at Shanghai Jiao Tong University, China, from 2018 to 2023, and joined Shanghai Maritime University in 2023. His research interests include marine new energy, energy storage, shipping power technology, carbon capture, and porous media. He has published about eighty papers in international journals. He is currently an associate editor and editorial board member of the *Journal of Porous Media* and *Special Topics & Reviews in Porous Media*.

Dr. Chen Yang is a professor at the College of Chemical Engineering, Fuzhou University, China. He received his Ph.D. in 2012 from University of Shizuoka, Japan. From 2012 to 2014, he worked as a postdoc at the National Polytechnic Institute of Toulouse (INPT), France. He joined Fuzhou University, China, as an associate professor in 2014 and was promoted to full professor in 2016. His research activity covers various areas, such as transport theory in porous media and chemical process intensification. He has published more than seventy journal papers. Due to the achievement of catalytic packing design, Dr. Yang received several awards, including first prize for S&T Progress Award (2016, China Petroleum and Chemical Industry Association; 2018, People's Government of Fujian Province).

Liwei Zhang is a professor at the Institute of Rock and Soil Mechanics, Chinese Academy of Sciences. He obtained his Ph.D. from Carnegie Mellon University, USA, in 2013. Prof. Zhang's research has been focused on experimental and numerical studies on the reactive transport of fluid in porous media, mineral dissolution, and precipitation processes in the subsurface, and the evolution of the pore structure of cementitious materials in corrosive environment. His research activities have resulted in 2 books, 5 book chapters, 16 patents, and more than 130 journal articles and conference proceedings. He was selected as a recipient of the GSE Emerging Scientist Award in 2023.

Contents

Preface

Porous media plays a pivotal role in science and engineering. They have unique structural characteristics that endow them with extensive application value, spanning from underground water reservoirs to energy conversion devices, and from biomedical fields to environmental engineering applications. Porous media possess widely distributed pore structures, which can not only store fluids and gases but also influence the processes of mass and heat transfer. Therefore, a profound understanding of the characteristics and behaviors of porous media is crucial for optimizing their application performance and addressing relevant engineering problems.

Transport Perspectives for Porous Medium Applications presents a comprehensive exploration of the diverse applications and advancements within the realm of porous media. This book delves into the intricate interplay between transport phenomena and porous structures across various fields, offering insights into the fundamental principles and practical implications of porous medium applications.

The ten chapters in this volume span a wide spectrum of topics, ranging from novel apparatus development for measuring relative permeability to the design and application of metal foam flow fields in proton exchange membrane (PEM) fuel cells. Each chapter encapsulates a distinct facet of porous media research, showcasing the breadth and depth of its interdisciplinary relevance.

One of the notable features of this book is its focus on cutting-edge research methodologies and technologies, such as microstructure reconstruction within porous electrodes of PEM fuel cells and the utilization of hierarchical porous catalysts for enhancing mass transfer and coke resistance in dry reforming of methane (DRM) processes. By elucidating these advanced techniques, this book aims to foster innovation and inspire further exploration in the field of porous medium applications.

Furthermore, this volume offers a forward-looking perspective on emerging trends and future directions in porous media research. From the evolution of pore structures induced by CO_2 carbonation in cement and concrete to the transport mechanisms in membranes used for desalination applications, the chapters provide valuable insights into the ongoing evolution of porous medium technologies and their potential impact on diverse industries.

As editors, we express our gratitude to the chapter authors for their insightful contributions and dedication to advancing porous medium applications. This book is a valuable resource for researchers, engineers, and students, fostering dialogue and collaboration across disciplines to drive innovation in transport phenomena within porous media. It offers a systematic understanding of porous media applications while encouraging cross-fertilization and innovative development in the field. We hope this book inspires researchers to delve deeper into porous media, collectively exploring uncharted territories and making significant contributions to global challenges in energy, environment, and engineering.

Huijin Xu
Professor,
Department of Thermal Energy,
Shanghai Maritime University,
Shanghai, China

Chen Yang
Professor,
College of Chemical Engineering,
Fuzhou University,
Fuzhou, China

Liwei Zhang
Professor,
Chinese Academy of Sciences,
Institute of Rock and Soil Mechanics,
Wuhan, China

Chapter 1

The Development and Application of Novel Apparatus for Relative Permeability Measurement

Shaicheng Shen, Zhiming Fang and Xiaochun Li

Abstract

Relative permeability plays a crucial role in understanding the characteristics of gas and water seepage in porous media and in establishing production schedules in practical engineering applications. However, accurately determining water saturation in the relative permeability measurement is challenging due to the minimal detectable amount of movable water in ultra-low-permeability rocks. This chapter introduces a novel method to determine water production during relative permeability measurement. We developed an apparatus that combines a separator with a high-precision differential pressure transducer (DPT) to measure the variation in water production during the experiment. The repeatability of measurements using this apparatus was tested, and the results demonstrated high stability. In addition, we used this apparatus to investigate the gas–water relative permeability in high-rank coal. The results indicate that the effect of displacement pressure on the relative permeability properties depended on the type of gas. The carbon dioxide–water system exhibited a significantly larger two-phase flow span compared to the helium–water or methane–water system at the same displacement pressure. Moreover, the relative permeability of the carbon dioxide–water system showed a higher sensitivity to the displacement pressure compared to the helium–water and methane–water systems.

Keywords: porous media, relative permeability, water saturation, differential pressure transducer (DPT), ultra-low-permeability rocks

1. Introduction

To address the global challenge of climate change, many researchers have shifted their focus toward unconventional clean energy extraction [1] and carbon dioxide geological sequestration technologies [2–4]. These technologies aim to mitigate carbon emissions or enhance clean energy production by injecting carbon dioxide into various subsurface formations such as saline aquifers, coal beds, and shale formations. As these reservoirs typically contain both liquid and gas phases, it is essential to understand the multiphase flow characteristics in reservoirs for effective project assessment. Relative permeability plays a vital role in characterizing the flow potential of each phase within the reservoir. It represents the ratio of the effective permeability of each phase (gas or liquid) to the absolute permeability of the reservoir. This parameter is crucial for reservoir simulation, where accurate modeling of fluid flow behavior is necessary for optimizing carbon dioxide storage or enhanced energy production. By quantifying the relative permeability of different phases, researchers can better understand how fluids move and interact within the reservoir. This knowledge is essential for predicting the performance and efficiency of carbon dioxide sequestration or energy extraction projects.

In the measurement of relative permeability using the unsteady state method [5, 6], accurate determination of water saturation was challenging for reservoirs with low porosity and permeability. This is because rocks with small porosity and high irreducible water saturation tend to displace only a small amount of water during the experiment. Over the past 50 years, researchers have developed various equipment and measurement techniques to address this challenge. One approach is to collect the displaced water in a graduated fluid receiver tube and estimate the volume based on the change in water level. Dabbous et al. [7] designed an apparatus specifically for measuring the relative permeability of coal using this method. Similarly, Durucan et al. [8] developed an apparatus for measuring the relative permeability, and the change in water volume was measured by conventional volumetric method. However, it is important to note that the gas–water interface during the experiment is unstable, which may affect the accuracy of data determination. Another widely used method is the gravimetric method [9–12], where the weight change of water in the separator is measured using an electronic balance. However, the presence of pipelines between the separator and the core holder can introduce volatility in the electronic balance readings. Nuclear magnetic resonance (NMR) relaxation has been employed by Sun et al. [13] to measure the gas–water relative permeability of coal. In this method, the water content in the pores is estimated using the total amplitude of T_2. Schembre and Kovscek [14] and Zhao et al. [15] have utilized X-ray CT measurement to calculate the relative permeability. Water saturation data is obtained from images of the core slices at a certain time. Both NMR and CT provide direct measurements of water saturation, avoiding potential errors caused by dead volumes in the apparatus. However, the limited resolution of NMR and CT imposes constraints on the sample size used in the experiment, which may affect the representativeness of the results. Additionally, the high cost of NMR and CT poses a significant barrier to their widespread adoption.

In this chapter, a novel apparatus for measuring the gas–liquid relative permeability of ultra-low-permeability rocks was introduced. This apparatus combined a specially engineered separator and a high-precious DPT in the relative permeability measurement system. The separator was designed to withstand high-pressure conditions, ensuring the simulation of a real reservoir environment. The DPT, on the other hand, was capable of monitoring the water content in real time, providing continuous and precise data during the experiment. Importantly, the DPT offered a more cost-effective solution for relative permeability measurement compared to NMR and CT. To validate the performance of the apparatus, thorough verification tests were conducted. These tests aimed to assess the accuracy, reliability, and stability of the measurements obtained using the apparatus. Additionally, the impact of displacement pressure and gas type on the gas–water relative permeability of coal was investigated. The findings from these experiments contribute to a better understanding of fluid flow behavior in ultra-low-permeability rocks.

2. Apparatus

2.1 The design of relative permeability apparatus

We developed a novel apparatus designed to accurately measure the gas–water relative permeability of ultra-low-permeability rocks. The schematic diagram of the apparatus can be seen in **Figure 1(a)**. The apparatus is composed of five main components: a gas injection and control system, a core holder system, a separation system, s back-pressure control system, and a data acquisition system (DAQ). Based on the schematic diagram of the novel apparatus, we constructed the relative permeability device, as shown in **Figure 1(b)**. The gas injection and control system comprises a gas source and a syringe pump. The syringe pump can withstand high pressure, with a maximum capacity of 25.86 MPa. This system allows precise control over the gas

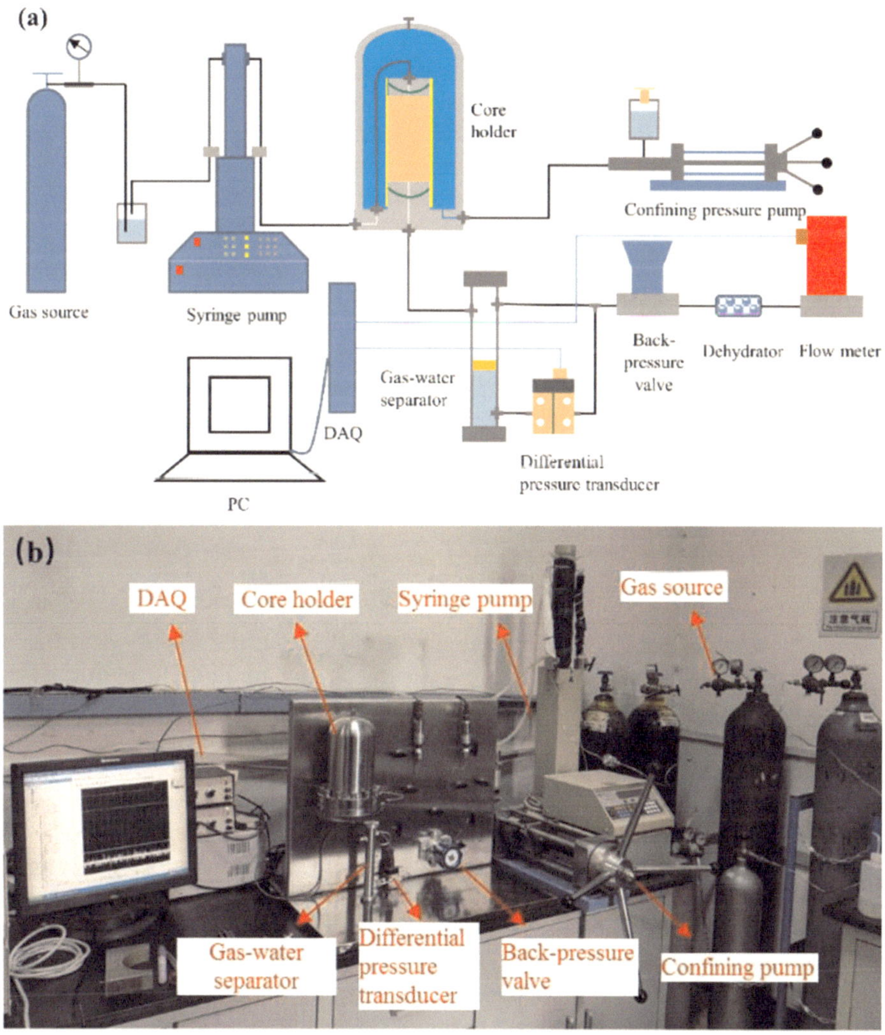

Figure 1.
The diagram of the novel apparatus [16]. (a) the structure diagram of the relative permeability measurement system; and (b) picture of the apparatus for relative permeability measurement.

injection process, ensuring accurate measurements. Within the core holder system is a core holder itself, along with a confining pump. The core holder guarantees secure placement of the rock sample, and the confining pump applies a maximum confining pressure of 25 MPa. The separation system comprises a gas–water separator and a DPT. The back-pressure control system incorporates a back-pressure valve and a flow meter. The back-pressure valve serves the purpose of maintaining a consistent back-pressure during the experiment, while the flow meter, with a measurement range of 50 ml/min and an accuracy of ±0.5%RD + ±0.1%FS, accurately measures the flow rate. For effective data acquisition, the DAQ system was equipped with various pressure sensors and a data acquisition card. This configuration allows simultaneous data collection from each sensor, thereby enhancing the reliability of the acquired data.

The characteristic of our device lies in two aspects: the design of the gas–water separation system and the arrangement of the downstream pipeline. The gas–water separation system in our apparatus consists of a cylindrical separator and a high-precision differential pressure transducer (DPT). The separator was specifically designed to efficiently collect the displaced water, while the DPT accurately monitors changes in the water level, enabling precise determination of water production. To prevent water evaporation and minimize system errors, a small amount of oil was added to the separator. This ensures that the collected water remains intact throughout the measurement process. Furthermore, the arrangement of the pipelines connecting the core holder and the separator plays a crucial role in reducing residual water in the pipelines. By arranging these pipelines vertically, we can effectively minimize the amount of residual water, enhancing the accuracy of the measurements.

One key advantage of our device is its adjustability in terms of measurement range and accuracy. The measurement accuracy of the displaced water is influenced by both the inner diameter of the separator and the precision of the DPT. As a result, we have the flexibility to design separators with different diameters to meet specific measurement requirements. This adjustability allows for precise and tailored measurements based on the unique characteristics of different rock samples. The relationship between the differential pressure and the displaced water volume follows the liquid pressure formula:

$$\Delta p = \rho_w g \Delta h = \rho_w g \frac{4\Delta V}{\pi d^2} \tag{1}$$

where Δp is the change of differential pressure, Pa; ρ_w is the density of water, kg/m^3; g is the acceleration of gravity, m/s^2; Δh is the change of liquid level in the separator, m; ΔV is the change of water volume in the separator, m^3; d is the inner diameter of the separator, m.

In theory, it is indeed possible to enhance the measurement accuracy by reducing the inner diameter of the separator indefinitely, as demonstrated by Eq. (1). However, it is important to consider practical limitations and avoid potential capillary effects. We recommend a minimum inner diameter of 5 mm for the separator. In our specific study, we utilized a separator with an inner diameter of 6 mm. To ensure accurate measurements, we employed a high-precision DPT with a measurement range of 1.4 kPa and an accuracy of ±0.25 FS. This DPT provides reliable and precise data that is essential for calculating the relative permeability. For researchers conducting relative permeability measurements using a similar device, we advise designing a specialized separation system to meet their specific requirements. This customization allows researchers to optimize the measurement accuracy and range based on their unique experimental conditions and objectives.

2.2 The performance of the novel apparatus

Measurement repeatability is indeed a crucial requirement for experimental equipment, and we have conducted assessments to evaluate it for gas permeability measurements on different rock types, including sandstone and coal. Additionally, we have performed verification experiments to assess the repeatability of water production during relative permeability measurements.

To measure the gas permeability of sandstone, we employed the steady state method while maintaining a constant gas injection rate throughout the measurement process. The differential pressure between the upper and lower sides of the sample was recorded in **Figure 2(a)**. In Test 1, nitrogen gas was injected at a rate of 20 ml/min, and the differential pressure eventually stabilized at 30.330 kPa. In Tests 2 and 3, nitrogen gas was injected at a rate of 10 ml/min, resulting in coinciding differential pressure curves. The final differential pressure reached 14.548 kPa. Using Darcy's law, we calculated the gas permeability, as depicted in **Figure 2(b)**. The gas permeabilities were measured to be 11.20, 11.07, and 11.07 mD for Test 1, Test 2, and Test 3, respectively. The average gas permeability across the three tests was found to be 11.11 mD, with a corresponding standard deviation of 0.0613.

In our study, we employed the pulse-decay method to measure the permeability of coal. The decay curves of the differential pressure over time were plotted in **Figure 3(a)**. These curves exhibited an exponential decay pattern. Using the fitted results obtained from the decay curves, we applied the equation proposed by Jones [17] to calculate the gas permeability of the coal sample. The gas permeabilities were determined to be 5.56, 5.49, and 5.45 μD for Test 1, Test 2, and Test 3, respectively. These values are presented in **Figure 3(b)**. The average gas permeability across the three tests was found to be 5.5 μD, with a corresponding standard deviation of 0.0437. This indicates that the measurements are consistent and repeatable, as evidenced by the small standard deviation.

After verifying the repeatability of gas permeability, we conducted nitrogen-water relative permeability experiments on sandstone using the unsteady state method. The sandstone samples were subjected to a confining pressure of 8.5 MPa, with the downstream connected to the atmosphere. To maintain consistency, the

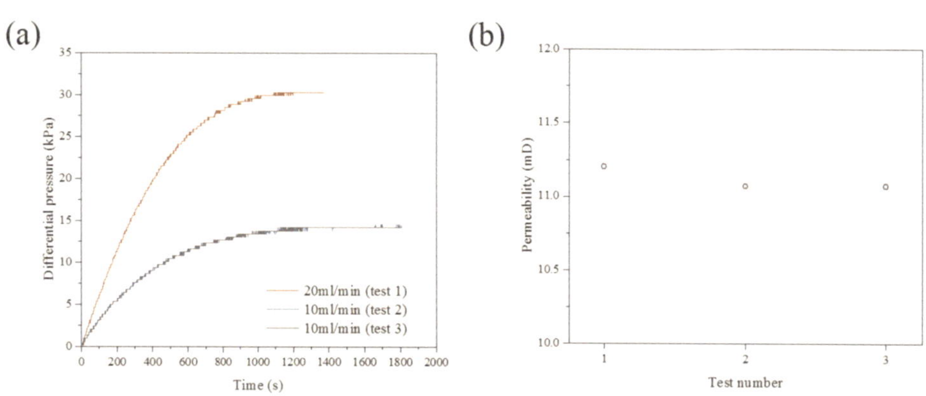

Figure 2.
The measurement results of gas permeability of sandstone by the steady-state method. (a) the differential pressure curves; and (b) the measured gas permeability.

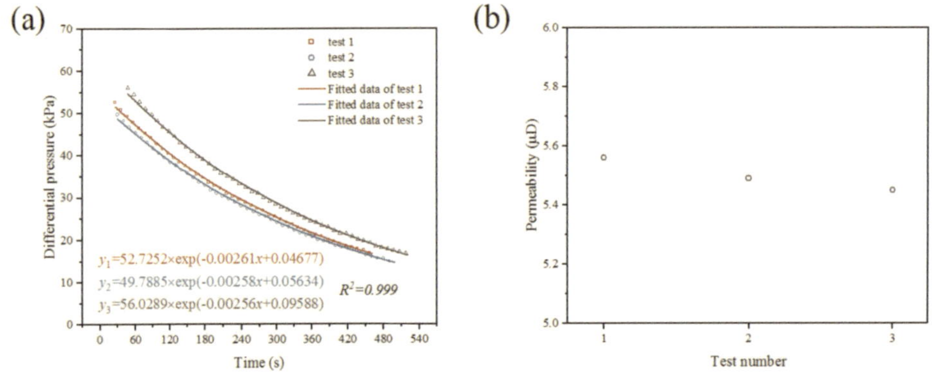

Figure 3.
The measurement results of gas permeability of coal by the pulse-decay method. (a) the differential pressure curves; and (b) the measured gas permeability.

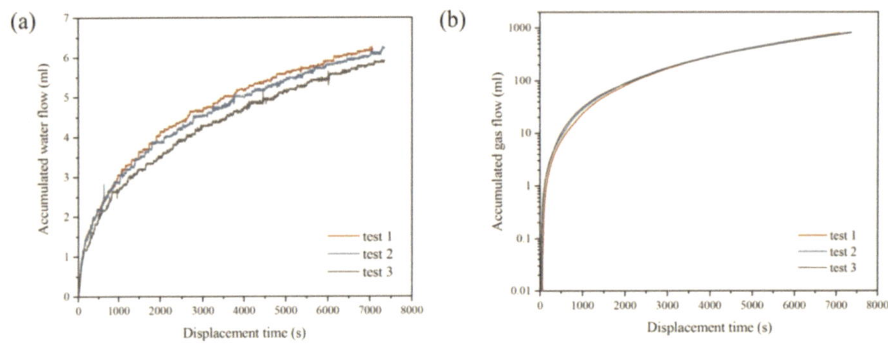

Figure 4.
The water and gas production during the relative permeability measurement [16]. (a) Water production; and (b) gas production.

displacement pressure was kept at 0.24 MPa throughout the experiments. The relative permeability experiment was repeated three times, and the water and gas production data obtained from each repetition are presented in **Figure 4**. The data demonstrate good agreement across the three repetitions, indicating repeatability and reliability in the measurements. To quantitatively evaluate the differences between the three measurements, we calculated the global coefficient of variation (GCV) using the following formula:

$$GCV = \frac{1}{n}\sum_{i=1}^{n} \frac{\sqrt{\frac{1}{m}\sum_{j=1}^{m}\left(y_i^j - \overline{y_i}\right)^2}}{\overline{y_i}} \tag{2}$$

where n represent the time. j represents the test number. y_i^j represents the j th measurement result at the time i. $\overline{y_i}$ represents the average data at the time i.

The GCV for water production was 0.05, and for gas production was 0.06. The results obtained for both gas permeability and gas–water relative permeability, along with the stability observed in the measurements, further confirm the repeatability of the experimental apparatus used in this study.

3. Application

3.1 Experimental design and process

In this section, we utilized the previously evaluated novel apparatus to investigate the impact of displacement pressure and gas type on gas–water flow characteristics in coal. Coal is known for its unique swelling behavior induced by gas adsorption [18–21], and the extent of swelling is influenced by gas pressure [22–25]. Therefore, studying the effect of gas type and gas pressure on gas–water seepage behavior in coal is a logical step. The experiments used helium, methane, and carbon dioxide as the injected gases. The experimental conditions are summarized in **Table 1**. The injection pressure (P_{in}) ranged from 1.9 to 3.9 MPa, while the outstream pressure (P_{out}) remained constant.

The anthracite coal sample used in this investigation was collected from SiHe Coal Mine in the Qinshui Basin. The vitrinite reflectance of the coal sample was measured to be 3.37%. Prior to measuring the relative permeability, the coal sample was dried at 60°C for 24 hours to remove residual water. Subsequently, the gas permeability was measured using the pulse-decay method with helium. The confining pressure and pore pressure were set at 7.9 MPa and 2 MPa, respectively. After the gas permeability measurement, the coal sample was saturated with water for 3 days. Once complete water saturation was achieved, the measurement of relative permeability was conducted. Water production was determined by the DPT, while gas production was recorded by a gas flow meter. The relative permeability of gas and water was calculated based on the recorded data, using the calculation method described in our previous study [26]. By investigating the gas–water flow characteristics in coal under different gas types and pressures, this study aims to provide insights into the behavior of coal reservoirs and enhance our understanding of gas and water migration in such porous media.

3.2 Results of the displacement experiment

The gas permeability of the dried coal, measured with helium, was found to be 18.26 μD. **Figure 5** illustrates the recorded gas and water production for different gas–water systems at varying displacement pressures. For the helium–water and methane–water systems, it was observed that the total accumulated water flow remained below 1 ml, and the effect of displacement pressure on this flow was negligible. In the case of the helium–water system, the total accumulated water flow ranged from 0.238 ml to 0.327 ml as the displacement pressures varied. Similarly, for the methane–water system, the total accumulated water flow ranged from 0.179 ml to 0.387 ml with increasing displacement pressures. However, for the carbon dioxide–water system, the total accumulated water flow exceeded 1 ml, and it exhibited a significant increase with increasing displacement pressures. Specifically, as the displacement pressures increased from 1.8 MPa to 3.8 MPa, the total accumulated water flow increased from 1.488 ml to 5.178 ml.

Gas type	P_{in}/MPa	P_{out}/MPa	Effective stress/MPa
He, CH_4, CO_2	1.9, 2.9, 3.9	0.1	5.9

Table 1.
The experimental conditions of relative permeability measurement.

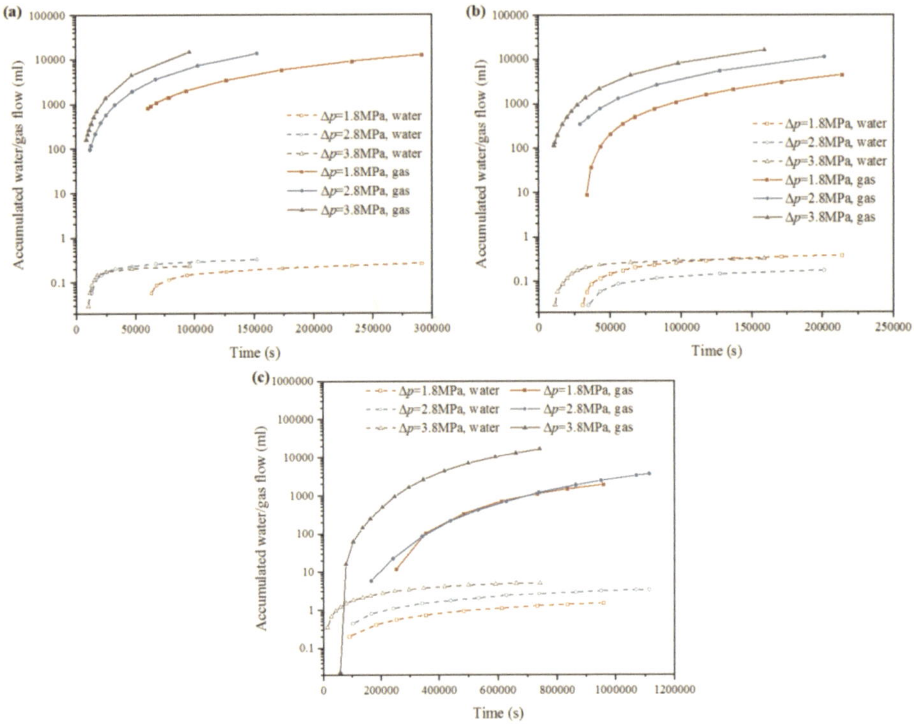

Figure 5.
The accumulated water/gas flow at different displacement pressures for different gas–water systems. (a) Helium–water system; (b) methane–water system; and (c) carbon dioxide–water system. Reprinted (adapted) with permission from reference [26]. Copyright 2023 American Chemical Society.

Based on the analysis of accumulated water/gas flow, we were able to calculate the gas–water relative permeability. The results of the helium–water relative permeability at different displacement pressures are presented in **Table 2**. It was observed that the irreducible water saturation was extremely high, ranging from 90.2% to 92.1%, indicating a narrow two-phase flow span for the helium–water system. Additionally, the maximum relative permeability of gas was found to be significantly higher than that of water. Specifically, the maximum relative permeability of gas ranged from 13.97% to 15.48%, while the maximum relative permeability of water was much lower, ranging between 1.82% and 2.77%. It should be noted that the influence of displacement pressures on the characteristic parameters of relative permeability was not apparent. Across different displacement pressures, the irreducible water saturation remained relatively consistent, ranging from 90.2% to 92.1%. Similarly, the gas relative permeability at the irreducible water saturation varied only slightly, ranging from 13.97% to 15.48%.

The results of methane–water relative permeability under different displacement pressures are presented in **Table 3**. The irreducible water saturation ranged from 90.7% to 91.8%. This indicates a narrow two-phase flow span, similar to what was observed in the helium–water system. However, it is worth noting that the relative permeability of methane at the irreducible water saturation is significantly lower compared to helium. The methane relative permeability ranged from 4.67% to 5.26%, indicating that methane exhibited lower ability to flow through the coal matrix than helium. Moreover, the impact of displacement pressures on the relative permeability of methane–water showed similar trends to the helium–water system.

Δp = 1.8 MPa			Δp = 2.8 MPa			Δp = 3.8 MPa		
S_w/%	k_{rw}/%	k_{rg}/%	S_w/%	k_{rw}/%	k_{rg}/%	S_w/%	k_{rw}/%	k_{rg}/%
95.2	2.70	7.97	94.6	5.45	2.77	95.4	1.82	3.26
94.3	1.79	8.12	93.8	3.48	3.05	95.0	1.75	3.23
93.9	0.62	8.58	93.0	1.04	3.84	94.6	1.38	3.51
93.5	0.32	9.39	92.6	0.59	4.76	94.2	1.00	4.12
93.1	0.19	10.50	92.2	0.48	5.72	93.8	0.87	4.72
92.7	0.10	11.95	91.8	0.26	7.01	93.4	0.55	5.63
92.3	0.08	13.50	91.4	0.15	8.76	93.0	0.18	7.90
91.9	0.067	14.85	91.0	0.10	10.94	92.5	0.07	11.77
91.5	0.066	15.48	90.6	0.06	13.67	92.1	0.04	13.97
			90.2	0.05	15.21			

Reprinted (adapted) with permission from reference [26]. Copyright 2023 American Chemical Society.

Table 2.
Relative permeability measurement results for helium–water system.

Δp = 1.8 MPa			Δp = 2.8 MPa			Δp = 3.8 MPa		
S_w/%	k_{rw}/%	k_{rg}/%	S_w/%	k_{rw}/%	k_{rg}/%	S_w/%	k_{rw}/%	k_{rg}/%
97.0	2.40	0.00	93.1	0.46	1.72	96.3	3.14	1.25
96.6	1.93	0.24	92.7	0.39	2.03	95.9	2.12	1.32
96.2	1.34	0.89	92.3	0.26	2.55	95.4	0.84	1.51
95.8	0.91	1.46	91.9	0.15	3.06	95.0	0.60	1.78
95.4	0.58	1.83	91.5	0.08	3.82	94.6	0.63	2.09
95.0	0.50	2.20	91.1	0.05	4.77	94.2	0.55	2.37
94.6	0.46	2.46	90.7	0.04	5.26	93.8	0.42	2.69
94.2	0.38	2.70				93.4	0.27	3.09
93.8	0.28	2.95				93.0	0.15	3.57
93.3	0.22	3.24				92.6	0.07	4.11
92.9	0.19	3.59				92.2	0.05	4.77
92.5	0.16	3.97				91.8	0.03	5.15
92.1	0.10	4.42						
91.7	0.09	4.67						

Reprinted (adapted) with permission from reference [26]. Copyright 2023 American Chemical Society.

Table 3.
Relative permeability measurement results for methane–water system.

The results of carbon dioxide–water relative permeability under different displacement pressures are listed in **Table 4**. The irreducible water saturation in the carbon dioxide–water system ranged from 25.7% to 78.7% as the displacement

Δp = 1.8 MPa			Δp = 2.8 MPa			Δp = 3.8 MPa		
S_w/%	k_{rw}/%	k_{rg}/%	S_w/%	k_{rw}/%	k_{rg}/%	S_w/%	k_{rw}/%	k_{rg}/%
97.0	0.81	0.00	100.0	0.82	0.00	90.2	1.40	0.00
94.0	0.29	0.02	93.6	0.66	0.00	86.3	1.05	0.00
91.9	0.26	0.10	88.5	0.42	0.01	82.9	1.02	0.02
89.3	0.22	0.26	84.2	0.34	0.04	78.2	0.98	0.07
86.3	0.18	0.45	78.7	0.29	0.09	73.5	0.70	0.11
84.2	0.18	0.61	74.8	0.26	0.15	69.7	0.61	0.17
81.6	0.18	0.66	70.5	0.30	0.22	65.8	0.64	0.26
79.9	0.12	0.69	65.4	0.27	0.33	60.3	0.58	0.43
78.7	0.10	0.71	61.6	0.20	0.45	55.2	0.50	0.65
			57.7	0.22	0.54	50.5	0.46	0.87
			54.3	0.19	0.61	45.3	0.41	1.15
			51.8	0.15	0.65	39.8	0.32	1.45
			50.5	0.18	0.68	34.7	0.25	1.70
						30.4	0.19	1.92
						27.8	0.15	2.13
						25.7	0.12	2.22

Table 4.
Relative permeability measurement results for carbon dioxide–water system.

pressure increased from 1.8 MPa to 3.8 MPa. This indicates a wider range of two-phase flow span compared to the helium–water and methane–water systems. Additionally, the carbon dioxide relative permeability at the irreducible water saturation was found to be lower than that of helium and methane. Despite the lower irreducible water saturation in the carbon dioxide–water system, carbon dioxide exhibited lower ability to flow through the coal matrix compared to helium and methane. Unlike the helium–water and methane–water systems, the characteristics of carbon dioxide–water relative permeability were significantly influenced by the displacement pressure. As the displacement pressure increased from 1.8 MPa to 3.8 MPa, the irreducible water saturation decreased from 78.7% to 25.7%. These findings demonstrate that the behavior of carbon dioxide–water relative permeability differs from that of helium–water and methane–water systems. The wider range of irreducible water saturations and the influence of displacement pressure highlight the unique characteristics of carbon dioxide as a displacing fluid in coal.

3.3 Impact of displacement pressure and gas type on gas–water relative permeability

The relative permeability curves of different gas–water systems at various displacement pressures are illustrated in **Figure 6**. It was observed that the gas type significantly affected the relative permeability of anthracite coal. For the helium–

water and methane–water systems, the shapes of the gas–water relative permeability curves were almost identical, while curves for the carbon dioxide–water system were flatter. Furthermore, the carbon dioxide–water system exhibited a substantial increase in the span of the two-phase flow.

The effect of gas type and displacement pressure on the characteristic parameters of relative permeability is presented in **Figure 7**. It was observed that the variation of irreducible water saturation with displacement pressure is influenced by the gas type. In the case of the helium–water and methane–water systems, there was little change in irreducible water saturation with the increase in displacement pressure. This finding was consistence with the study conducted by Yang et al. [27], who examined the effect of gas (N_2) displacement pressures on the displaced water content in low-permeability coal. They found that the irreducible water content stabilized when the displacement pressure exceeded 2 MPa. However, for the carbon dioxide–water system, the irreducible water saturation showed a significant and rapid decrease as the displacement pressure increased from 1.8 MPa to 3.8 MPa. This indicates that the impact of displacement pressure on irreducible water saturation is more pronounced in the carbon dioxide–water system compared to the helium–water and methane–water systems. The difference in irreducible water saturation with displacement pressure can be attributed to the variation in the interfacial tension between different

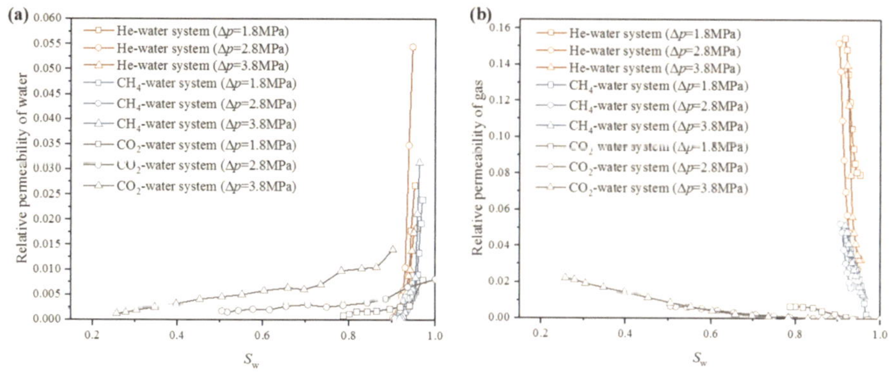

Figure 6.
The effect of displacement pressure on gas–water relative permeability curves. (a) the relative permeability of water; and (b) the relative permeability of gas. Reprinted (adapted) with permission from reference [26]. Copyright 2023 American Chemical Society.

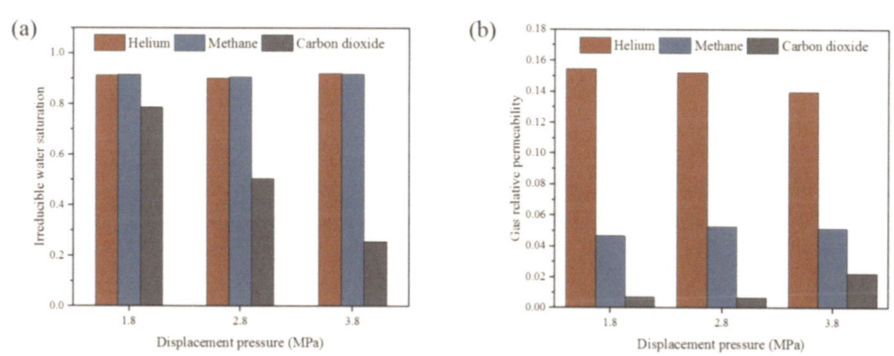

Figure 7.
The characteristic parameters of relative permeability curve at different displacement pressures. (a) Irreducible water saturation; and (b) gas relative permeability at irreducible water saturation.

gas–water systems. Previous studies [28–30] have indicated that the interfacial tension between helium and water, as well as methane and water, remains relatively close within the pressure range of 0.1–4 MPa (approximately 66–73 mN/m). However, the interfacial tension between carbon dioxide and water decreased rapidly, from 72 to 46.8 mN/m [31, 32], as the pressure increased to 4 MPa. The decline in interfacial tension promotes water displacement in the experiment. Additionally, the dissolved carbon dioxide alters the contact angles between carbon dioxide, water, and coal [33], resulting in a shift in coal wettability from water-wet to carbon dioxide wet [34]. The change in wettability enhanced water recovery efficiency during the drainage process, leading to a decrease in irreducible water saturation in the carbon dioxide–water system with increasing displacement pressure.

Furthermore, it was observed that the gas permeability at irreducible water saturation exhibited a more pronounced variation with gas type rather than displacement pressure. The sequence of swelling induced by gas adsorption in coal is widely recognized as $CO_2 > CH_4 > He$ [35]. Consequently, the gas permeability in the helium–water, methane–water, and carbon dioxide–water systems followed the same sequence.

4. Limitations

Precision determining the relation between the accumulated gas/water flow and water saturation is essential in the relative permeability experiment. However, experimental errors were present in this study. The accumulated gas/water curves indicated that the gas preferentially penetrated the sample during the drainage process for measuring helium–water and methane–water relative permeability. It was recognized as an illusion caused by errors in the experimental system. During the drainage process, the displaced water initially accumulates on the sample outlet surface and then fills the dead volume, which is unavoidable in the device design. The accumulated water overcomes the adhesion between the water and the surface of the coal sample and flows into the separator due to gravity and pressure gradient. This leads to the illusion that an early gas breakthrough occurs in relative permeability measurement. Furthermore, it is well-known that the absolute permeability of coal is influenced by gas adsorption [21]. Relative permeability is defined as the ratio of effective permeability to absolute permeability. Therefore, the results of relative permeability are affected by the choice of absolute permeability. In this study, the absolute permeability used for the calculations was based on He permeability. As a result, the calculated relative permeability values obtained in the experiment were extremely small.

5. Conclusion

In this chapter, the development of a novel apparatus for measuring gas–water relative permeability in ultra-low-permeability rocks was presented. The apparatus utilized high-precision DPT to determine the water saturation of the rock samples. This combination of DPT and separator allows the relative permeability measuring device to adjust the water measurement range and measurement accuracy according to specific requirements. To ensure the reliability of the measurement results, gas permeability tests were conducted three times on sandstone and coal samples, and the relative permeability of sandstone was also tested three times. The repeatability of the measurements was confirmed, providing confidence in the accuracy and consistency of the obtained data. The investigation then focused on studying the impact of displacement pressures and gas types on the relative permeability of anthracite coal. The findings and conclusions drawn from this study can be summarized as follows:

1. The combination of DPT and a separator is a feasible method for determining the water saturation of ultra-low-permeability rocks during relative permeability measurements.

2. The gas permeability and relative permeability measurements exhibit good repeatability, with low standard deviations and GCV values, indicating the accuracy and consistency of the obtained data.

3. The impact of displacement pressure on relative permeability properties varies depending on the gas type. Helium–water and methane–water systems show no significant differences in two-phase flow span with increasing in displacement pressure, while the carbon dioxide–water system exhibits an increase in two-phase flow span.

4. Gas type plays a significant role in influencing the relative permeability of coal. Carbon dioxide–water relative permeability demonstrates a substantial two-phase flow span compared to helium–water or methane–water systems under the same displacement pressure.

Acknowledgements

This work was supported by the Central Guidance on Local Science and Technology Development Fund of Inner Mongolia Autonomous Region (No.2022ZY0018), the National Key Research and Development Program of China (No.2018YFB0605601).

Conflict of interest

The authors declare that they have no conflicts of interest.

Notes/thanks

The authors would like to express their sincere gratitude to the American Chemical Society (ACS) and the Institute of Physics (Great Britain) for granting permission to utilize the figures and tables presented in this chapter. Their generosity allows us to enhance the quality and comprehensiveness of our research findings. We are thankful for their support in disseminating scientific knowledge and promoting collaboration in the academic community.

Author details

Shaicheng Shen[1,2], Zhiming Fang[1*] and Xiaochun Li[1]

1 State Key Laboratory of Geomechanics and Geotechnical Engineering, Institute of Rock and Soil Mechanics, Chinese Academy of Sciences, Wuhan, Hubei, China

2 University of Chinese Academy of Sciences, Beijing, China

*Address all correspondence to: zmfang@whrsm.ac.cn

References

[1] Puri R, Yee D, editors. Enhanced coalbed methane recovery. In: SPE Annual Technical Conference and Exhibition. Society of Petroleum Engineers; 1990

[2] Bachu S. Sequestration of CO_2 in geological media in response to climate change: Road map for site selection using the transform of the geological space into the CO_2 phase space. Energy Conversion and Management. 2002;**43**(1):87-102. DOI: 10.1016/S0196-8904(01)00009-7

[3] Smit B. Carbon capture and storage: Introductory lecture. Faraday Discussions. 2016;**192**:9-25. DOI: 10.1039/c6fd00148c

[4] Mikunda T, Brunner L, Skylogianni E, Monteiro J, Rycroft L, Kemper J. Carbon capture and storage and the sustainable development goals. International Journal of Greenhouse Gas Control. 2021;**108**:103318. DOI: 10.1016/j.ijggc.2021.103318

[5] Honarpour M, Mahmood SM. Relative-permeability measurements - An overview. Journal of Petroleum Technology. 1988;**40**(8):963-966. DOI: 10.2118/18565-Pa

[6] Shen SC, Fang ZM, Li XC. Laboratory measurements of the relative permeability of coal: A review. Energies. 2020;**13**(21). DOI: 10.3390/en13215568

[7] Dabbous MK, Reznik AA, Taber JJ, Fulton PF. The permeability of coal to gas and water. Society of Petroleum Engineers Journal. 1974;**14**(06):563-572. DOI: 10.2118/4711-a

[8] Durucan S, Ahsan M, Shi JQ, Syed A, Korre A. Two phase relative permeabilities for gas and water in selected European coals. Fuel. 2014;**134**: 226-236. DOI: 10.1016/j.fuel.2014.05.040

[9] Gash BW. Measurement of "rock properties" in coal for coalbed methane production. In: Paper Presented at the SPE Annual Technical Conference and Exhibition, Dallas, Texas. Society of Petroleum Engineers. 1991. pp. 221-230

[10] Shen J, Qin Y, Li YP, Wang G. Experimental investigation into the relative permeability of gas and water in low-rank coal. Journal of Petroleum Science and Engineering. 2019;**175**: 303-316. DOI: 10.1016/j.petrol.2018.12.041

[11] Shaw D, Mostaghimi P, Armstrong RT. The dynamic behaviour of coal relative permeability curves. Fuel. 2019;**253**:293-304. DOI: 10.1016/j.fuel.2019.04.107

[12] Zhang Z, Yan D, Yang S, Zhuang X, Li G, Wang G, et al. Experimental studies on the movable-water saturations of different-scale pores and relative permeability of low-medium rank coals from the southern Junggar Basin. Journal of Natural Gas Science and Engineering. 2020;**83**:103585. DOI: 10.1016/j.jngse.2020.103585

[13] Sun X, Yao Y, Ripepi N, Liu D. A novel method for gas-water relative permeability measurement of coal using NMR relaxation. Transport in Porous Media. 2018;**124**(1):73-90. DOI: 10.1007/s11242-018-1053-y

[14] Schembre JM, Kovscek AR. A technique for measuring two-phase relative permeability in porous media via X-ray CT measurements. Journal of Petroleum Science and Engineering. 2003;**39**(1-2):159-174. DOI: 10.1016/S0920-4105(03)00046-9

[15] Zhao HP, Hu JH, Wang JK, Zhang Y. A comprehensive model for calculating relative permeability based on spontaneous imbibition and CT scanning measurement. Fuel. 2019;**247**:287-293. DOI: 10.1016/j.fuel.2019.03.056

[16] Shen S, Fang Z, Li X. A novel method for determining water saturation of porous media in relative permeability measurement using DPT. Measurement Science and Technology. 2023;**34**(3): 035901. DOI: 10.1088/1361-6501/aca2cd

[17] Jones SC. A technique for faster pulse-decay permeability measurements in tight rocks. SPE Formation Evaluation. 1997;**12**(1):19-25. DOI: 10.2118/28450-Pa

[18] Robertson EP, Christiansen RL. Measurement of Sorption-Induced Strain. Idaho Falls, United States: Idaho National Lab.(INL); 2005 Report No.: INL/CON-05-001830532

[19] Karacan CO. Heterogeneous sorption and swelling in a confined and stressed coal during CO_2 injection. Energy & Fuels. 2003;**17**(6):1595-1608. DOI: 10.1021/ef0301349

[20] Pan Z, Connell LD. A theoretical model for gas adsorption-induced coal swelling. International Journal of Coal Geology. 2007;**69**(4):243-252

[21] Pan ZJ, Connell LD, Camilleri M. Laboratory characterisation of coal reservoir permeability for primary and enhanced coalbed methane recovery. International Journal of Coal Geology. 2010;**82**(3-4):252-261. DOI: 10.1016/j.coal.2009.10.019

[22] Dutta P, Bhowmik S, Das S. Methane and carbon dioxide sorption on a set of coals from India. International Journal of Coal Geology. 2011;**85**(3-4):289-299. DOI: 10.1016/j.coal.2010.12.004

[23] Song Y, Jiang B, Lan FJ. Competitive adsorption of $CO_2/N_2/CH_4$ onto coal vitrinite macromolecular: Effects of electrostatic interactions and oxygen functionalities. Fuel. 2019;**235**:23-38. DOI: 10.1016/j.fuel.2018.07.087

[24] Saurabh S, Harpalani S. Anisotropy of coal at various scales and its variation with sorption. International Journal of Coal Geology. 2019;**201**:14-25. DOI: 10.1016/j.coal.2018.11.008

[25] Zhao Y, Cui D, Liu J, Wei M, Liu Y. Evolution of coal permeability under constant effective stresses: Direct measurements and numerical Modeling. Energy & Fuels. 2021;**35**(19):15489-15501. DOI: 10.1021/acs.energyfuels.1c01425

[26] Shen S, Fang Z, Li X, Xue Q, Ci H. Experiment on the gas permeability and gas–water relative permeability in high-rank coal with a self-developed device. Energy & Fuels. 2023;**37**(3):2201-2212. DOI: 10.1021/acs.energyfuels.2c04052

[27] Yang MF, Yang ZB, Sun B, Zhang ZG, Liu HL, Zhao JL. A study on the flowability of gas displacing water in low-permeability coal reservoir based on NMR technology. Frontiers of Earth Science. 2020;**14**(4):673-683. DOI: 10.1007/s11707-020-0837-x

[28] Massoudi R, King AD. Effect of pressure in the surface tension of water. Adsorption of low molecular weight gases on water at 25 degree. Journal of Physical Chemistry. 1974;**78**(22): 2262-2266. DOI: 10.1021/j100615a017

[29] Sun C, Chen G, Yang L-y. Interfacial tension of methane + water with surfactant near the hydrate formation conditions. Journal of Chemical & Engineering Data. 2004;**49**:1023-1025. DOI: 10.1021/JE049948P

[30] Yasuda K, Mori YH, Ohmura R. Interfacial tension measurements in water–methane system at temperatures from 278.15 K to 298.15 K and pressures up to 10 MPa. Fluid Phase Equilibria. 2016;**413**:170-175. DOI: 10.1016/j.fluid.2015.10.006

[31] Chun BS, Wilkinson GT. Interfacial tension in high-pressure carbon dioxide mixtures. Industrial and Engineering Chemistry Research. 1995;**34**:4371-4377. DOI: 10.1021/ie00039a029

[32] Hebach A, Oberhof A, Dahmen N, Kogel A, Ederer H, Dinjus E. Interfacial tension at elevated pressures - Measurements and correlations in the water plus carbon dioxide system. Journal of Chemical and Engineering Data. 2002; **47**(6):1540-1546. DOI: 10.1021/je025569p

[33] Han FS, Busch A, van Wageningen N, Yang JL, Liu ZY, Krooss BM. Experimental study of gas and water transport processes in the inter-cleat (matrix) system of coal: Anthracite from Qinshui Basin, China. International Journal of Coal Geology. 2010;**81**(2):128-138. DOI: 10.1016/j.coal.2009.11.001

[34] Siemons N, Bruining H, Castelijns H, Wolf K-H. Pressure dependence of the contact angle in a CO_2–H_2O–coal system. Journal of Colloid and Interface Science. 2006;**297**(2):755-761. DOI: 10.1016/j.jcis.2005.11.047

[35] Gensterblum Y, Ghanizadeh A, Krooss BM. Gas permeability measurements on Australian subbituminous coals: Fluid dynamic and poroelastic aspects. Journal of Natural Gas Science and Engineering. 2014;**19**: 202-214. DOI: 10.1016/j.jngse.2014.04.016

Chapter 2

The Elastic Properties and Yield Strengths of Low-Density Honeycombs and Open-Cell Foams

Hanxing Zhu

Abstract

This chapter aims to briefly review the main theoretical and finite element simulation results on the elastic properties and yield strengths of regular hexagonal honeycombs, Kelvin open cell foams, random irregular Voronoi honeycombs and open-cell foams, and discuss about their deformations mechanisms. The book chapter further introduces the effects of other parameters such as the degree of cell regularity, imperfection of defects, cell size, solid volume fraction and material distribution, on the elastic properties and yield strengths of these cellular materials. Voronoi honeycombs and open cell foams can be related to their mechanical properties.

Keywords: honeycombs, open-cell foams, elastic properties, yield strengths, size effects

1. Introduction

Living natural materials are usually cellular materials, examples of which include bones and wood. Such structures tend to use as little as possible solid material to provide sufficient mechanical properties (e.g., strength, stiffness, and energy absorption capacity) and the desired physical or biological functions. In general, open-celled cellular materials can be classified into two types. The first type is honeycombs which are often treated as 2D materials in some studies, and the other type is open cell foams which are 3D materials and are not suitable to be treated as 2D materials.

In practical applications, there are many different types of regular or irregular honeycombs, open-cell foams and two-dimensional (2D) or three-dimensional (3D) auxetic cellular materials. Numerous publications are relevant to elastic buckling, plastic collapse and dynamic performance, which are involved in geometric nonlinearity, material nonlinearity and strain rate effects. In most engineering applications, the strength and stiffness are the most important mechanical properties of cellular materials, which are essential to ensure the structural integrity and to enable the relevant physical functions. Due to the page limit, this chapter only briefly summarizes the key theoretical and simulation research results on the static strength and stiffness of low-density regular hexagonal honeycombs, Kelvin open-cell foams and their random irregular Voronoi structures, and then briefly introduces the effects of some key factors on these properties. Sections 2 and 3 present the elastic properties for the first type of open-cell cellular materials, i.e. honeycombs; Sections 4 and 5 present the these properties for open-celled foams.

The honeycomb and open-cell foam structures could be used under different environmental conditions such as varying and extreme temperature and humidity. In these cases, as long as the honeycombs and open-cell foams are initially sufficiently strong and stiff, their long-term structural integrity and mechanical performance mainly depend on the solid material from which the honeycombs and the open-cell foams are made.

2. The elastic properties of honeycombs

There are different types of regular honeycombs, for example, triangular honeycombs and square honeycombs. In most relevant research works, honeycombs are treated as 2D materials. To partition a two-dimensional domain into a cellular structure with n-identical cells, a regular hexagonal honeycomb has the smallest total cell wall length. Many people have studied the in-plane elastic properties of regular hexagonal honeycombs with a low relative density and a uniform cell wall thickness [1–4]. For low-density regular hexagonal honeycombs with a uniform cell wall thickness t and a cell wall length l, their relative density is $\rho = \frac{2t}{\sqrt{3}l}$. Their in-plane elastic properties are isotropic, their in-plane Poisson's ratio is approximately 1.0 and their in-plane Young's modulus is given as [1, 3].

$$E_1 = 1.5E_S\rho^3 \tag{1}$$

where E_S is the Young's modulus of the solid material from which the honeycombs are made.

Regular hexagonal honeycombs at different size scales have five independent elastic constants and the analytical results of all these elastic constants are obtained [4]. It has been found that when the cell wall thickness is small, these properties are size-dependent due to the strain gradient effects at the micro-meter scale and the surface effects at the nanometer scale [4]. In general, the smaller the cell size, the stiffer the honeycombs. When the cell wall thickness is at the millimeter scale or larger, the elastic constants of regular hexagonal honeycombs depend only on E_S and ρ, and are independent of the cell size or wall thickness [1–4]. It is noted in [4] that when the out-of-plane dimension w (i.e., the width of the cell walls) of low-density hexagonal honeycombs is much larger than the cell wall thickness t, their in-plane Young's modulus should be $E_1 = 1.5E_S\rho^3/(1 - v_S^2)$, instead of that given by Eq. (1), where v_S is the Poisson's ratio of the solid material from which the honeycombs are made.

Silva et al. [5] used finite element method to simulate the elastic properties of non-periodic Voronoi honeycombs. Zhu et al., for the first time, defined the degree of cell regularity for random irregular Voronoi honeycombs, and studied the effects of cell regularity on the elastic properties of periodic Voronoi honeycombs [6] and related their geometric properties of the irregular Voronoi cells [7] to the elastic properties of irregular honeycombs. To partition an area A into n-identical hexagonal cells, the distance between the cell centers of any two neighboring cells is

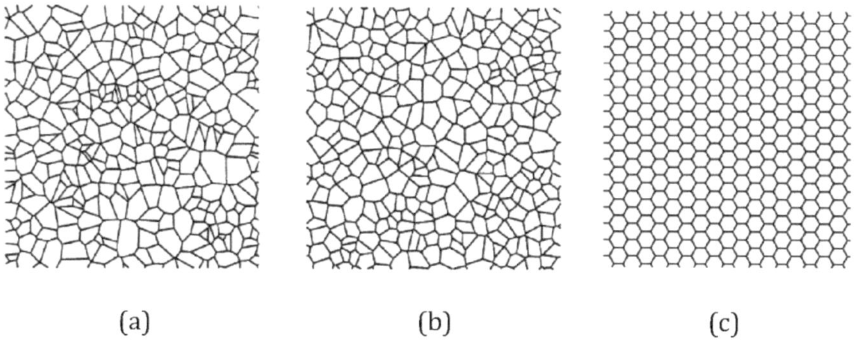

Figure 1.
Periodic honeycombs with 300 complete cells [6], (a) $\alpha = 0.1$, (b) $\alpha = 0.4$ and (c) $\alpha = 1.0$.

$$d_0 = \left(\frac{2A}{n\sqrt{3}}\right)^{1/2} \tag{2}$$

To partition the same area into n random Voronoi cells, the minimum distance between the centers of any two neighboring cells can be obtained as δ. Obviously, δ should be smaller than d_0, otherwise, it is impossible to construct n random Voronoi cells. The degree of cell regularity is thus defined in [6, 7] as

$$\alpha = \delta / d_0 \tag{3}$$

When δ is 0, the corresponding honeycomb is a fully random Voronoi honeycomb; when δ equals d_0, the corresponding honeycomb is a perfectly regular hexagonal honeycomb [6, 7]. **Figure 1** shows random Voronoi honeycombs with $\alpha = 0.1$ and $\alpha = 0.4$, and a perfectly regular hexagonal honeycomb with $\alpha = 1.0$ [6].

Zhu et al. [6] used finite element method (ABAQUS Timoshenko beam elements) and periodic random representative volume element (RVE) models, as shown in **Figure 1**, to study the effects of relative density and cell regularity on the in-plane

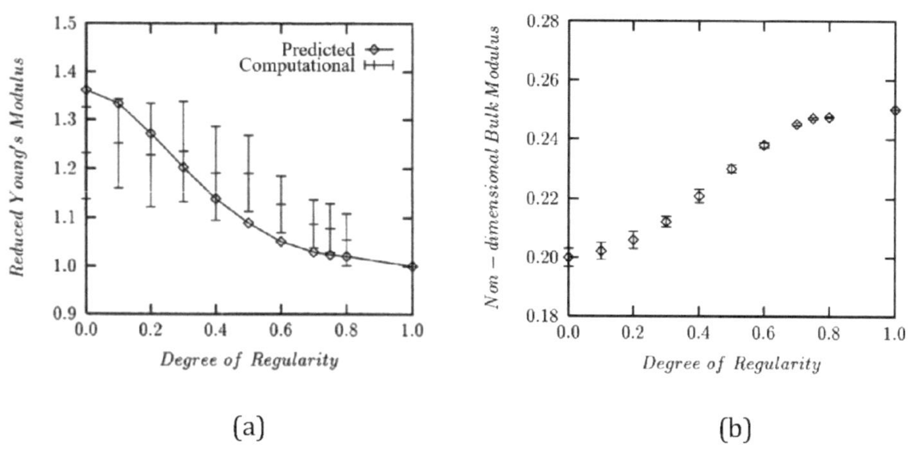

Figure 2.
Effects of cell irregularity on the in-plane Young's modulus (a) and bulk modulus (b) of random irregular Voronoi honeycombs with $\rho = 0.01$ [6].

elastic properties (i.e., Young's modulus, shear modulus, Poisson's ratio, and bulk modulus) of low-density random Voronoi honeycombs. They found that the elastic properties are in-plane isotropic. The obtained effects of cell regularity on the in-plane Young's modulus and bulk modulus of Voronoi honeycombs with $\rho = 0.01$ are presented in **Figure 2**, where the Young's modulus has been normalized by $1.5\,E_S\rho^3$ or $1.5\,E_S\rho^3/(1-v_S^2)$ depending on the value of t/w, and the bulk modulus has been normalized by ρE_S. In addition, they [6] developed a springs-in-parallel model to relate the geometric properties [7] to the elastic properties of random irregular Voronoi honeycombs, and the predicted results agree very well with the finite element computational results, as can be seen in **Figure 2(a)**. The results in **Figure 2** show clearly that the smaller the degree of cell regularity, the larger the in-plane Young's modulus and the smaller the in-plane bulk modulus of the random irregular Voronoi honeycombs.

It is noted that due to the effects of axial compression and transverse shear of the cell walls, the normalized Young's moduli and shear moduli of both regular hexagonal honeycombs and Voronoi honeycombs decrease with the increase of the relative density. This is because compared to the bending deformation of the cell walls, the deformations caused by the axial compression and transverse shear are very small and negligible when the honeycomb relative density is very small. However, they increase with the honeycomb relative density and could become comparable if the relative density is large, making the normalized Young's moduli and shear moduli smaller with the increase of the honeycomb relative density. Due to the strain gradient effect at the micro-scale and the surface effect at the nano-scale, the thinner the cell walls, the larger the in-plane Young's modulus of the regular or random irregular Voronoi honeycombs [4, 8]. For single-level nano-sized honeycombs or multi-level nano-structured hierarchical honeycombs, both their elastic properties and geometric properties could be controlled to vary over large ranges by applying an electric potential to change the surface stress of the nano-sized cell walls [4, 8]. At the micro- or macro-scale, the out-of-plane elastic constants of both regular hexagonal and random irregular Voronoi honeycombs with a uniform cell wall thickness are $E_3 = E_S\rho$, $G_{31} = \frac{1}{2}G_S\rho$ and $v_{31} = v_S$, where G_S is the shear modulus of the solid material from which the honeycombs are made.

For the in-plane uniaxial deformation of low-density regular and irregular honeycombs at different size scales, cell wall bending is always the dominant deformation mechanism. When a honeycomb undergoes in-plane deformation, the largest bending moment usually occurs at the cell wall junction. Thus, material distribution along the cell wall can significantly affect the in-plane Young's modulus of honeycombs. By properly increasing solid material along the cell wall from the middle to the junction, the in-plane Young's modulus can be remarkably increased [9, 10]. In addition, imperfections such as inclusions and holes can sharply reduce the in-plane Young's modulus of honeycombs [11].

3. The yield strength of honeycombs

Many people [3, 10–14] have studied the yield strength of regular hexagonal and random Voronoi honeycombs. For a low-density regular hexagonal honeycomb with a uniform cell wall thickness and a small ratio of w/t, the initial in-plane uniaxial yield strength can be obtained as [3].

$$\sigma_{1y}^{i} = \frac{1}{3}\rho^2\sigma_{ys} \tag{4}$$

If the same low-density regular hexagonal honeycomb is made of an elastic and perfectly plastic material, its in-plane full yield strength is 50% higher than the initial yield strength and given as [3].

$$\sigma_{1y}^{f} = \frac{1}{2}\rho^2\sigma_{ys} \tag{5}$$

In Eqs. (4) and (5), σ_{ys} is the yield strength of the solid material from which the honeycomb is made.

For a low-density regular hexagonal honeycomb with a uniform cell wall thickness and a large ratio of w/t, the uniaxial in-plane initial yield strength of the honeycomb is given as [10].

$$\sigma_{1y}^{i} = \frac{1}{3}\rho^2\sigma_{ys}/\sqrt{\left(1 - v_s + v_s^2\right)} \tag{6}$$

If the same low-density regular hexagonal honeycomb is made of an elastic and perfectly plastic material, the full yield strength can be obtained as [10].

$$\sigma_{1y}^{f} = \frac{1}{\sqrt{3}}\rho^2\sigma_{ys} \tag{7}$$

The in-plane initial or full shear yield strength of a low-density regular honeycomb can be obtained by dividing the corresponding in-plane initial or full uniaxial yield strength by $4/\sqrt{3}$. In general, random Voronoi honeycombs have a slightly lower in-plane uniaxial yield strength than a perfectly regular hexagonal honeycomb [12]. The in-plane yield strength of a perfectly regular hexagonal honeycomb is less sensitive to the in-plane hydrostatic stress than a random Voronoi honeycomb. The lower the degree of cell regularity, the smaller the in-plane uniaxial yield strength of random Voronoi honeycombs. When a metal honeycomb is in-plane crushed, 90% of the externally applied energy is absorbed by the plastic hinges [13]. Due to the strain gradient effect at the micro-scale and the surface effect at the nano-scale, micro- and nano-honeycombs have a larger in-plane yield strength than their macro-counterparts. It is noted that with the increase of the relative density, the in-plane yield strength of regular hexagonal honeycombs becomes increasingly smaller than those given by Eqs. (4)–(7) because of the effects of axial compression and transverse shear of the cell walls [10]. Properly allocating the solid material along the cell walls could significantly enhance the in-plane yield strength of honeycombs [10, 12], and imperfection such as holes and missing cell walls could cause sharp decrease in the in-plane yield strength of honeycombs [11, 12, 14].

When a regular hexagonal honeycomb made of an elastic and perfectly plastic material is under out-of-plane uniaxial tension or compression (i.e., under uniaxial tension or compression in direction 3), its initial plastic yield strength is the same as the full yield strength, and is given as $\sigma_{3y}^{i} = \sigma_{3y}^{f} = \rho\sigma_{ys}$. When the regular hexagonal honeycomb is under out-of-plane pure shear in the direction parallel to some of the cell walls, the plastic shear yield strength is $\sigma_{31y} = \frac{1}{2}\rho\sigma_{ys}$. When the same regular hexagonal honeycomb is under out-of-plane pure shear in the direction perpendicular to some of the cell walls, the plastic shear yield strength is slightly higher and given as $\sigma_{32y} = \frac{1}{\sqrt{3}}\rho\sigma_{ys}$. Thus, the out-of-plane shear yield strengths of a perfectly regular hexagonal honeycomb are different in different directions. For a Voronoi honeycomb with the same relative density and uniform cell wall thickness, its out-of-plane uniaxial tensile or compressive yield strength is the same as that of a regular hexagonal honeycomb, its out-of-plane shear yield strength is between $\frac{1}{2}\rho\sigma_{ys}$ and $\frac{1}{\sqrt{3}}\rho\sigma_{ys}$. Nonuniform distribution of the solid material along the cell wall length reduces the out-of-plane shear yield strength of the honeycombs and does not change the out-of-plane tensile/compressive yield strength. Nano-sized honeycombs could have larger out-of-plane tensile/compressive and shear yield strengths compared to their macro-sized counterparts with the same relative density.

It is noted that the results on the in-plane and out-of-plane elastic buckling, plastic collapse and dynamic behaviors are not included here due to the page limit.

4. The elastic properties of open-cell foams

Open cell foams are another type of cellular materials, and their typical geometric structure can be seen in **Figure 3(a)** [15]. Taking strut bending as the sole deformation mechanism and using dimensional analysis, Gibson and Ashby [3, 18] have obtained the Young's modulus and shear modulus for low-density three-dimensional open-cell foams and their results are given as

$$E = C_1\rho^2 E_S \text{ and } G = C_2\rho^2 E_S \tag{8}$$

In Eq. (8), ρ is the relative density of the open-cell foams, C_1 and C_2 are constants depending on the strut cross-sectional shape. In addition, they [3, 18] have experimentally measured the Young's modulus, shear modulus and Poisson's ratio of low-density open-cell foams, and found that their elastic properties are almost isotropic, i.e., satisfy the relation $E = 2(1 + v)G$. By fitting the analytical results to the experimentally measured data of low-density open-cell foams, they [3, 18] found that $C_1 = 1$, $C_2 = 3/8$ and $v = 1/3$.

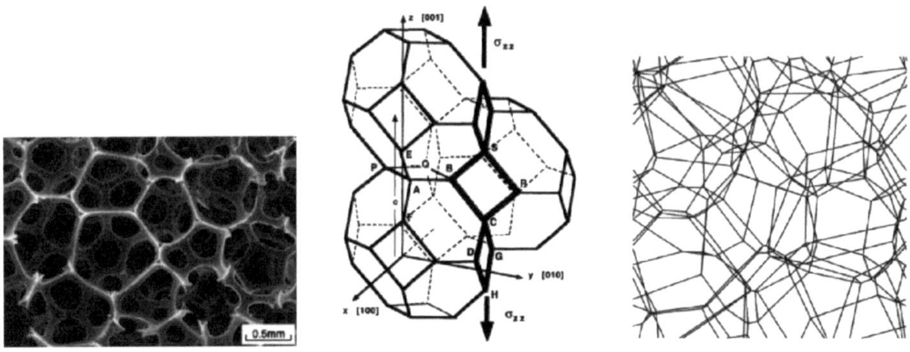

(a) (b) (c)

Figure 3.
Geometric structure and models of three-dimensional open-cell foams, (a) the actual structure of an open cell foam [15], (b) tetrakaidecahedral (BCC or Kelvin) cell model [16], (c) a periodic random Voronoi model with 27 complete cells [17].

Periodic unit cell structures or periodic representative volume elements (RVEs) are often used to study the effective mechanical properties of cellular materials. Taking strut bending, torsion and axial stretching/compression as the deformation mechanisms, Zhu et al. [16] have used the tetrakaidecahedron (also called BCC or Kelvin foam) structural model, as shown in **Figure 3(b)**, to analyze the elastic properties of low-density open-cell foams, obtained the general expressions of the Young's modulus E_1, shear modulus G_{12}, Poisson's ratio υ_{12} as well as the bulk modulus K, and found that the elastic properties obtained from the tetrakaidecahedron model are almost isotropic. According to the manufacturing process of polymer open-cell foams, each of cell struts is shared by the three neighboring cells, thus the shape of the cell strut cross-section is a plateau border. When the cell strut cross-section is a plateau border, the Young's modulus, shear modulus, Poisson's ratio and the bulk modulus of the Kelvin open-cell foam are predicted as [16]

$$E_1 = \frac{1.009\rho^2 E_S}{1 + 1.514\rho} \tag{9}$$

$$G_{12} \approx 0.33\rho^2 E_S \tag{10}$$

$$\upsilon_{12} = \frac{1}{2} \cdot \frac{1 - 1.514\rho}{1 + 1.514\rho} \tag{11}$$

$$K = \frac{1}{9}\rho E_S \tag{12}$$

Where the foam relative density $\rho = \frac{3A}{2\sqrt{2}l^2}$, l and A are the strut length and strut cross-sectional area of the tetrakaidekahedral cell.

The theoretical Young's modulus given by Eq. (9) is exactly the same as Gibson and Ashby's experimental result $E = \rho^2 E_S$ measured from low-density open-cell foams [3, 18]. As the tetrakaidecahedron structure has cubic symmetry, it has only three independent elastic constants, i.e., E_1, G_{12} and υ_{12}. For low-density Kelvin open-cell foams, the Zener's anisotropy factor is [16].

$$A^* = \frac{2(S_{11} - S_{12})}{S_{44}} = \frac{2(1 + \upsilon_{12})G_{12}}{E_1} \approx 0.98 \tag{13}$$

Thus, the elastic properties of the low-density tetrakaidecahedron (or BCC or Kelvin) open-cell foams are almost isotropic (i.e., $E_1 \approx 2(1 + \upsilon_{12})G_{12}$). To partition a space into n-identical cells, the tetrakaidecahedral cell has the smallest cell surface area. If all the cell surfaces have the same thickness and the same tensile stress during the foam manufacturing process, the foam structural system has the lowest energy. Thus, the tetrakaidecahedron is the best regular cell structural model for analyzing the mechanical properties of open-cell foams because of the two reasons: the mechanical isotropy and the lowest energy state.

A real open-cell foam usually has a random irregular geometric structure. Thus, a 3D random Voronoi representative volume element (RVE) can more realistically reflect the geometric structure and the mechanical properties of open-cell foams. Zhu et al., for the first time, defined the degree of regularity for random irregular Voronoi open-cell foams, and studied the effects of cell regularity on the elastic properties of periodic Voronoi open-cell foams [17] and related their geometric properties [19] to their elastic properties. To partition a space V into n-identical tetrakaidecahedral cells, the minimum distance d_0 between any two neighboring cell centers is [17],

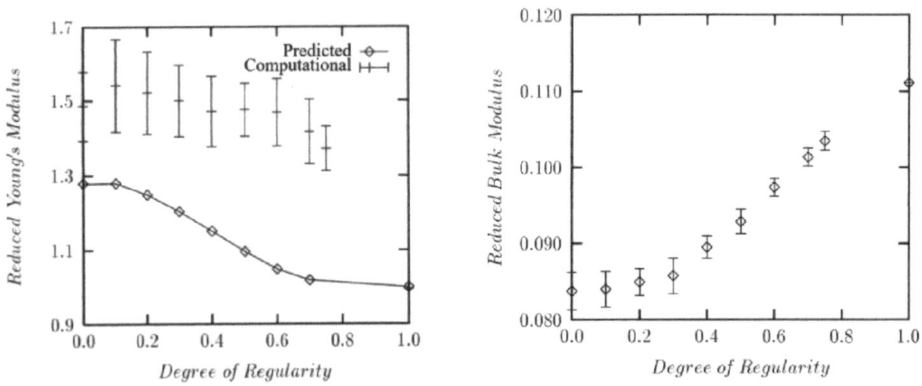

(a)

(b)

Figure 4.
Effects of cell regularity on the (a) Young's modulus and (b) bulk modulus of random Voronoi open-cell foams with $\rho = 0.01$ *[17].*

$$d_0 = \frac{\sqrt{6}}{2}\left(\frac{V}{\sqrt{2}n}\right)^{\frac{1}{3}} \tag{14}$$

To construct a random irregular Voronoi tessellation with n cells in space V, the distance δ between the centers of any two neighboring cells must be smaller than d_0; otherwise, it is impossible to construct n cells. Thus, the regularity degree of a 3D Voronoi tessellation is defined as [17, 19],

$$\alpha = \frac{\delta}{d_0} \tag{15}$$

For a regular lattice with identical tetrakaidecahedral cells, δ equals d_0 and $\alpha = 1$. For a completely random irregular Voronoi tessellation, δ equals 0 and $\alpha = 0$.

Zhu et al. [17] developed a computer code to construct periodic 3D random irregular Voronoi open-cell foams as shown in **Figure 3(c)**, and then used ABAQUS Timoshenko beam elements and periodic boundary conditions to study the effects of relative density and cell regularity on the Young's modulus, shear modulus, Poisson's ratio and bulk modulus of Voronoi open-cell foams. **Figure 4** shows the effects of cell regularity on the Young's modulus and bulk modulus of low-density open-cell foams with $\rho = 0.01$, where the Young's modulus has been normalized by $\rho^2 E_S$ and the bulk modulus has been normalized by ρE_S. As can been seen from **Figure 4**, low-density irregular random Voronoi open-cell foams have a much larger Young's modulus and much smaller bulk modulus than those of a perfect Kelvin foam; the lower the degree of cell regularity, the larger the normalized Young's modulus (**Figure 4(a)**) and the smaller the normalized bulk modulus (**Figure 4(b)**). The elastic properties of low-density Voronoi open-cell foams are isotropic, and their Poisson's ratios are very close to 0.5 [17]. Zhu et al. [17] found that both the Poisson's ratio and the normalized Young's modulus of Voronoi open-cell foam reduce significantly with the increase of the foam's relative density due to the effects of the axial compression and transverse shear of the cell struts, and the tangent modulus of Voronoi open-cell foams decreases substantially with the increase of the foam compressive strain. Gibson and Ashby [3, 18] have experimentally found that the normalized Young's moduli of random irregular open-cell foams are close to 1.0 and in general much smaller than those given in **Figure 4(a)**. The reasons may be the effects of the foam relative density (i.e., $\rho > 0.01$), the boundary condition and the compressive strain in the experimental measurements. It is noted that Christenson [20] and Warren and Kraynik [21] have analyzed the bulk modulus for low-density open-cell foams and obtained a result of $\rho E_S/9$, which is the same as that of a perfect Kelvin foam [16], but larger than those of random irregular Voronoi open-cell foams, as can be seen in **Figure 4(b)**.

Zhu et al. [17] also developed a springs-in-parallel model to relate the geometric properties [19] to the elastic properties of 3D random irregular Voronoi open-cell foams, and to qualitatively predict how cell regularity affects the Young's modulus of the random irregular Voronoi open-cell foams, as shown by the predicted results in **Figure 4(a)**. Although the springs-in-parallel model could well predict the trend of the cell regularity effects, the predicted Young's moduli exhibit large deviations from the exact results obtained from the finite element simulations. This model is just an attempt to qualitatively explain the reason why an irregular open-cell foam has a larger Young's modulus than a more regular open-cell foam. So far, there is no available analytical model that could accurately predict the effects of cell regularity on the elastic properties of irregular open-cell foams.

For low-density open-cell foams with $\rho < 0.1$, strut bending and torsion are the dominant deformation mechanisms, although strut axial compression and transverse shear play an important role in the deformation and can significantly affect the elastic properties [16, 17]. As the largest bending moment usually occurs at the strut junctions, allocating more solid material near the strut junctions could remarkably increase the Young's modulus of low-density open-cell foams [22–24]. In addition, defects could significantly reduce the stiffness of open-cell foams [25]. Due to the strain gradient effect at the micro-scale and the surface effect at the nano-scale, for open-cell foams with the same relative density, the thinner the cell struts, the larger the Young's modulus and shear modulus [26, 27]. For single-level nano-sized open-cell foams or multi-level nano-structured hierarchical open-cell foams, both their elastic properties and geometric properties could be controlled to vary over large ranges by applying an electric potential to change the surface stress of the nano-sized cell struts [27].

5. The yield strength of open-cell foams

Many people [3, 18, 22, 23, 25, 28–30] have studied the plastic collapse strength of different types of open-cell foams. For low-density open-cell foams, the plastic collapse strength is given by [3, 18].

$$\sigma_y = 0.3\rho^2\sigma_{ys} \tag{16}$$

or

$$\sigma_y = 0.23\rho^{3/2}\left(1+\rho^{1/2}\right)\sigma_{ys} \tag{17}$$

As strut bending and torsion are the dominant deformation mechanisms and the largest bending moment usually occurs at the strut junctions, allocating more solid material near the strut junctions could increase the yield strength [22, 23]. In addition, defects could significantly reduce the yield strength of open-cell foams [25]. It is found that low-density random irregular Voronoi open-cell foams are stiffer and stronger than Kelvin open-cell foams for compressive strain up to 0.15 [31] and have a larger yield strength than Kelvin open-cell foams [22]. In addition, it has been experimentally found that for Au open-cell foams with the same relative density and the size of the cell struts at the nanometer scale, the smaller the cell struts, the larger the yield strength of the nano open-cell foams [32].

6. Conclusions

There are thousands of publications on the mechanical properties of honeycombs and open-cell foams. Theoretical analyzes are usually based on specific geometric models, and many different types of geometric models have been proposed and analyzed for honeycombs and open-cell foams. Regular hexagonal honeycomb, which has five independent elastic constants, is the best structural model for regular honeycombs [1, 3], and regular Kelvin foam, which has three independent elastic constants, is the best structural model for regular open-cell foams [16]. Natural honeycombs and open-cell foams, however, always have a certain degree of irregularity. Thus, random irregular 2D [6, 7] and 3D [17, 19, 31] Voronoi tessellations with different degrees of cell regularity are the most ideal geometric models for the analyzes of the mechanical properties of honeycombs and open-cell foams.

Due to the page limit, this chapter just briefly reviews/introduces the generally recognized knowledge about the elastic properties and yield strength of honeycombs and open-cell foams. The mechanical properties of both honeycombs and open-cell foams strongly depend on their relative density, and the larger the relative density, the stiffer and stronger the honeycombs and open-cell foams. In addition to the effects of relative density, their Young's modulus, shear modulus and yield strength are proportional to those of the solid material from which the honeycombs or open-cell foams are made. It has been found that defects could significantly reduce the strength and stiffness of honeycombs [11, 12] and open-cell foams [25].

For the in-plane deformation of low-density honeycombs, cell wall bending is the dominant deformation mechanism, although the axial compression and transverse shear of the cell walls also play an important role [1–6]. Low-density random irregular honeycombs have larger in-plane Young's modulus, smaller in-plane yield strength and bulk modulus, and the same out-of-plane Young's modulus and yield strength compared to the same density regular hexagonal honeycombs [6, 12]. For low-density open-cell foams, strut bending and torsion are the dominant deformation mechanisms, although the axial compression and transverse shear of the cell struts also play an important role [16, 17, 31]. In general, random irregular open-cell foams have larger Young's modulus and yield strength, and smaller bulk modulus than those of the same density regular Kelvin open-cell foam [16, 17, 31].

As the largest bending moment usually occurs at the cell wall or cell strut junctions, allocating more solid material near the cell wall or cell strut junctions could significantly increase the stiffness and yield strength of honeycombs and open-cell foams [9, 10, 12, 22–24]. Due to the strain gradient effects at the micro-scale and the surface effects at the nanometer scale, the thinner the cell walls or cell struts, the stiffer and stronger the honeycombs or open-cell foams with the same relative density [4, 8, 26, 27, 32]. For nano-sized single-level or nano-structured multi-level hierarchical honeycombs and open-cell foams, their mechanical and geometric properties could be controlled to vary over a large range by application of an electric potential [4, 8, 26, 27].

Author details

Hanxing Zhu
School of Engineering, Cardiff University, Cardiff, UK

*Address all correspondence to: zhuh3@cardiff.ac.uk

References

[1] Gibson LJ, Ashby MF, Schajer GS, Robertson CI. The mechanics of two-dimensional cellular materials. Proceeding of the Royal Society of London. 1982;**382**:25-42

[2] Warren WE, Kraynik AM. Foam mechanics: The linear elastic response of two-dimensional spatially periodic cellular materials. Mechanics of Materials. 1987;**6**:27-37

[3] Gibson LJ, Ashby MF. Cellular Solids: Structure and Properties. 2nd ed. Cambridge: Cambridge University Press; 1997

[4] Zhu HX. Size-dependent elastic properties of micro- and nano-honeycombs. Journal of the Mechanics and Physics of Solids. 2010;**58**:696-709

[5] Silva MJ, Hayes WC, Gibson LJ. The effects of non-periodic microstructure on the elastic properties of two-dimensional cellular solids. International Journal of Mechanical Sciences. 1995;**37**: 1161-1177

[6] Zhu HX, Hobdell JR, Windle AH. Effects of cell irregularity on the elastic properties of 2D Voronoi honeycombs. Journal of the Mechanics and Physics of Solids. 2001;**49**:857-870

[7] Zhu HX, Thorpe SM, Windle AH. The geometrical properties of irregular 2D Voronoi tessellations. Philosophical Magazine A. 2001;**81**:2765-2783

[8] Zhu HX, Zhang HC, You JF, Kennedy D, Wang ZB, Fan TX, et al. The elastic and geometrical properties of micro- and nano-structured hierarchical random irregular honeycombs. Journal of Materials Science. 2014;**49**:5690-5702

[9] Li K, Gao XL, Subhash G. Effects of cell shape and cell wall thickness variations on the elastic properties of two-dimensional cellular solids. International Journal of Solids and Structures. 2005;**42**:1777-1795

[10] Zhu HX, Chen CY. Combined effects of relative density and material distribution on the mechanical properties of metallic honeycombs. Mechanics of Materials. 2011;**43**:276-286

[11] Chen C, Lu TJ, Fleck NA. Effect of inclusions and holes on the stiffness and strength of honeycombs. International Journal of Mechanical Sciences. 2001; **V43**:487-504

[12] Chen C, Lu TJ, Fleck NA. Effect of imperfections on the yielding of two-dimensional foams. Journal of the Mechanics and Physics of Solids. 1999; **47**:2235-2272

[13] Zhu HX. Large deformation pure bending of an elastic plastic power-law-hardening wide plate: Analysis and application. International Journal of Mechanical Sciences. 2007;**49**:500-514

[14] Silva MJ, Gibson LJ. The effects of non-periodic microstructure and defects on the compressive strength of two-dimensional cellular solids. International Journal of Mechanical Sciences. 1997;**39**: 549-563

[15] Gao G, Qi M, Li Y. Random equilateral Kelvin open-cell foam microstructures: Cross-section shapes, compressive behavior, and isotropic characteristics. Journal of Cellular Plastics. 2018;**54**:53-72

[16] Zhu HX, Knott JF, Mills NJ. Analysis of the elastic properties of open-cell foams with tetrakaidecahedral cells. Journal of the Mechanics and Physics of Solids. 1997;**45**:319-343

[17] Zhu HX, Hobdell JR, Windle AH. The effects of cell irregularity on the elastic properties of open cell foams. Acta Materialia. 2000;**48**:4893-4900

[18] Gibson LJ, Ashby MF. The mechanics of three-dimensional cellular materials. Proceedings of the Royal Society A: Mathematical, Physical and Engineering Sciences. 1982;**382**:43-49

[19] Zhu HX, Zhang P, Balint D, Thorpe SM, Windle AH, Lin J. The effects of regularity on the geometrical properties of Voronoi tessellations. Physica A. 2014; **406**:42-58

[20] Christenson RM. Mechanics of low density materials. Journal of the Mechanics and Physics of Solids. 1986; **34**:563-578

[21] Warren WE, Kraynik AM. The linear elastic properties of open-cell foams. Journal of Applied Mechanics. 1988;**55**: 341-346

[22] Mills NJ. The high strain mechanical response of the wet Kelvin model for open-cell foams. International Journal of Solids and Structures. 2007;**44**:51-65

[23] Jang WY, Kyriakides S, Kraynik AM. On the compressive strength of open-cell metal foams with Kelvin and random cell structures. International Journal of Solids and Structures. 2010;**47**:2872-2883

[24] Zargarian A, Esfahanian M, Kadkhodapour J, Ziaei-Ra S. Effect of solid distribution on elastic properties of open-cell cellular solids using numerical and experimental methods. Journal of the Mechanical Behavior of Biomedical Materials. 2014;**37**:264-273

[25] Lu ZX, Liu Q, Huang JX. Analysis of defects on the compressive behaviors of open-cell foams. Materials Science and Engineering A. 2011;**530**:285-296

[26] Zhu HX. Size-dependent elastic properties of micro- and nano-open-celled foams. Structural Longevity. 2011; **5**:17-24

[27] Zhu HX, Wang ZB. Size-dependent and tunable elastic and geometric properties of hierarchical nano-porous materials. Science of Advanced Materials. 2013;**5**:677-686

[28] Deshpande VS, Fleck NA. Isotropic constitutive models for metallic foams. Journal of the Mechanics and Physics of Solids. 2000;**48**:1253-1283

[29] Amsterdam E, De Hosson JTM, Onck PR. On the plastic collapse stress of open-cell aluminum foam. Scripta Materilia. 2008;**59**:653-656

[30] Karamooz Ravari MR, Kadkhodaei M, Badrossamay M, Rezaei R. Numerical investigation on mechanical properties of cellular lattice structures fabricated by fused deposition modeling. International Journal of Mechanical Sciences. 2014;**88**: 154-161

[31] Zhu HX, Windle AH. Effects of cell irregularity on the high strain compression of open-cell foams. Acta Materialia. 2002;**50**:1041-1052

[32] Hodge AM, Biener J, Hayes JR, Bythrow PM, Volkert CA, Hamza AV. Scaling equation for yield strength of nanoporous open-cell foams. Acta Materilia. 2007;**55**:1343-1349

Chapter 3

Microstructure Reconstruction and Gas-Liquid Two-Phase Transport Mechanism within Porous Electrodes of PEM Fuel Cells

Yulin Wang and Haokai Xu

Abstract

The structure of porous media is composed of skeleton particles and pores. Its micropores and solid skeleton characteristics lead to the capillary fingering movement of fluid in its porous media driven by capillary pressure. Currently, the methods of constructing porous media are mainly random construction and multi-scale imaging construction. The porous structure constructed by these two methods can show the real microstructure characteristics. The research on multiphase flow in microporous structure mainly includes VOF, MC, LBM, and other methods. In this chapter, taking the classic porous structure of polymer electrolyte membrane (PEM) fuel cell gas diffusion layer (GDL) as an example, GDL porous microstructure is constructed through random algorithm, and multiphase LBM is used to study two-phase flow in porous media to explore the relationship between porous structure characteristics and multiphase flow transport.

Keywords: capillary pressure, porous media, PEM fuel cell, gas-liquid two-phase transport, LBM

1. Introduction

In the field of energy and environment, the flow, heat, and mass transfer processes occurring in porous media generally exist [1, 2], such as the migration and distribution of light non-aqueous phase liquid (LNAPL) [3] and heavy non-aqueous phase liquid (DNAPL) [4] in porous media soil, the heat and catalytic reaction process of porous media catalyst particles in chemical granular bed chemical reactors [5], gas-liquid electric thermal multiphysics simulation heat, and mass transfer process in fuel cell porous structure assembly [6–8]. The heat and mass transfer process in the above porous media is closely related to the energy and environment on which human beings rely for survival, and involves all aspects of human life and industrial generation. Therefore, the study of the transport process in porous media is of great significance for rational energy exploitation and utilization, energy conservation and emission reduction, and environmental protection [9, 10]. To gain a deeper understanding of the influence of porous media structure on its internal transport, this chapter will take the porous media structure of proton exchange membrane (PEM) fuel cell GDL as an example to explore the gas-liquid two-phase transport process inside it.

PEM fuel cells are a type of power generation equipment that differs from traditional fossil fuels. It has the advantages of environmental protection, high efficiency, and fast response [11], and can be widely applied in various fields [12]. It is considered one of the most promising energy technologies. Despite the rapid development of PEM fuel cell technology in recent years, its large-scale commercialization still faces great challenges. Improving the water management capability of PEM fuel cell and strengthening the drainage performance and anti-aging performance of each component are important methods to improve the overall performance of PEM fuel cell [13, 14]. As an important component of PEM fuel cell, the GDL has the following functions: removing liquid water from the fuel cell, providing a pathway for the transport of reactant gas to the catalyst layer, providing structural support for the membrane, and promoting the transfer of electrons and heat [15, 16]. GDL is usually composed of a certain amount of hydrophilic carbon fibers (with a contact angle of 50°). To obtain more hydrophobic surface properties (with a contact angle of 130°), hydrophobic treatment with polytetrafluoroethylene (PTFE) is usually used to obtain high drainage performance [17, 18]. Optimizing GDL structural parameters with reasonable PTFE content and distribution is of great significance for improving the overall performance of fuel cells [19].

Various experimental works have been conducted to investigate the effects of PTFE content and distribution on the drainage and performance of fuel cells [20, 21]. However, the complex structure makes the reaction transport process in porous media much more complex than in a single media. To detect the gas-liquid two-phase transport process under the microstructure of GDL and obtain the influence mechanism of microstructure on drainage performance, microscopic visualization techniques such as neutron imaging, scanning electron microscopy (SEM), and X-ray technology were introduced [22–24]. Meyer et al. [24] studied GDL with different PTFE contents through neutron imaging and X-ray computed tomography. X-ray CT experimental results showed that PTFE forms like fibers attached to GDL carbon fibers, and high PTFE content will become an obstacle to water transport in GDL. Yu et al. [25] degraded GDL through accelerated stress testing and observed it through SEM. They report that the loss of carbon fibers and PTFE after GDL degradation leads to a decrease in hydrophobicity, seriously affecting the mass transfer characteristics of GDL. However, due to the limited spatial or temporal resolution of the existing experimental techniques, it is still difficult to study the transport process in microscopic porous media, especially the dynamic process of multiphase flow.

The science of studying the fluid transport process in porous media is called mechanics of flow through porous media [26, 27]. At present, the study on meso- and micro-level by using pore scale numerical simulation method has become the frontier topic in the field of microporous media study in the world, such as MC [28], PNM [29] and LBM [30–32] methods. LBM is suitable for simulating gas-liquid two-phase flow in microporous media structure, especially for simulating the heat and mass transfer process of gas-liquid electricity thermal coupling multiphysics simulation in porous media structure components of PEM fuel cell. Ira et al. [33] used pseudo-potential LBM to simulate the transport of liquid water in the microporous layer and diffusion layer of PEM fuel cell. The results showed that the content of hydrophobic solid skeleton had a great impact on the transport of gas-liquid two-phase in its interior and pointed out that GDL porous media with a certain degree of hydrophilic hydrophobic mixed structure had better liquid water transport performance. Han et al. [34] simulated the liquid water transport in the porous layer of PEM fuel cell through three-dimensional two-phase LBM. The results show that the porosity, pore structure, and liquid water saturation level in the porous layer have a great impact on the liquid water transport, which has guiding significance for the design of porous layer porous media structure optimization. Jithin et al. [35] used multi-relaxation LBM to simulate the fluid, heat, and material transport and oxygen reduction reaction in the cathode porous material assembly of PEM fuel cell. The numerical simulation considered the permeability, porosity, effective diffusion coefficient, and other properties of porous media, and the results showed that the solid skeleton material properties of the porous layer had an important impact on its internal heat and mass transport. Zhou et al. [36] built the three-dimensional porous microstructure of the gas diffusion layer and microporous layer of PEM fuel cell, and used LBM to simulate the coupling process of gas-liquid two-phase transport and oxygen reaction transport in the surface cracks. The simulation results show that the number of cracks has a great impact on the two-phase transport in porous media structure. When the cracks increase, liquid water covers the porous structure interface, which is not conducive to the transmission of liquid water.

This chapter briefly introduces the method of constructing microporous media structure by random algorithm and simulates gas-liquid two-phase transport in porous structure by SC pseudo-potential multiphase LBM. Then, it enumerates the influence of structural characteristics of PEM fuel cell GDL microporous media on two-phase transport and discusses the intrinsic relationship between its structural parameters and two-phase transport.

2. Construction of microporous media structure by random algorithm

The structure of porous media constructed by random algorithm can fully characterize its microstructure characteristics. At the same time, the model reconstructed by this method is completely based on program language or modeling software reconstruction. All structural parameters are flexible and controllable, and the cost is low. Therefore, the random algorithm is widely used in the study of simulating the heat and mass transfer process inside the microporous media.

GDL is a porous media material formed by a certain number of carbon fibers interlaced into a single layer and compressed to form a hydrophilic microstructure (50° ~ 80°). To achieve more hydrophobic surface physical properties (100° ~ 150°), GDL is usually placed in a certain mass fraction of PTFE solution for hydrophobic treatment. From the specific parameters of porous media structure, GDL is about 200 μm thick and carbon fibers are 7 μm in diameter, and its internal pore diameter varies from tens of micrometers to hundreds of micrometers, with a porosity of 60 ~ 80%. This chapter takes carbon paper GDL as an example for reconstruction, and the idea of reconstructing GDL through random algorithms is as follows:

1. Generate randomly distributed straight lines in a plane with a specified area size to represent the centerline of carbon fibers;

2. Expand the fiber centerline in the first step into a cylinder in three dimensions, forming a single carbon fiber. The processing method is to determine whether the distance from the pore lattice around the centerline to the centerline is less than the radius of the carbon fiber. If so, the lattice will be transformed into a fiber lattice, otherwise, it will still remain a pore lattice;

3. Repeat the work of the first and second steps to generate a number of carbon fibers in the plane, and the carbon fibers are randomly interlaced until the porosity of the layer meets the requirements (only carbon fibers);

4. PTFE is randomly generated from the carbon fiber layer obtained in the third step of work, and the generation position is mainly at the position where the intersecting pores of carbon fibers and carbon fibers are small, until the PTFE content of the carbon fiber layer meets the requirements (the porosity of each layer of carbon fibers is ultimately determined by the PTFE content);

5. Repeat the first to fourth steps to generate the target number of carbon fiber layers and then overlay them to obtain the fiber structure of the target thickness.

Figure 1 shows the steps for randomly reconstructing the microstructure of GDL.

3. Theory and application analysis of lattice Boltzmann method

LBM has better numerical stability and structural diversity in the study of multiphase transport processes in porous media structures. Generally, LBM can be divided into single relaxation time and multiple relaxation time according to the collision operator. Because the single relaxation time collision operator scheme has the advantages of simplicity and low computational cost, it is widely used in porous media flow simulation. This chapter mainly introduces the principle of Shan-Chen (SC) pseudo potential multiphase LBM with SRT collision operator and provides a theoretical basis for the subsequent numerical simulation of gas-liquid two-phase transport in GDL microporous media structure.

3.1 LBM content and introduction

For component k, the distribution function evolution equation for SRT and D2Q9 formats is:

$$f_{\mathrm{i}}^{\mathrm{k}}(x + c_{\mathrm{i}}\Delta t, t + \Delta t) - f_{\mathrm{i}}^{\mathrm{k}}(x,t) = -\frac{1}{\tau}\left[f_{\mathrm{i}}^{\mathrm{k}}(x,t) - f_{\mathrm{i}}^{\mathrm{eq,k}}(x,t)\right] \tag{1}$$

Where $f_{\mathrm{i}}^{\mathrm{k}}(x,t)$ is the density distribution function of component k (oxygen and liquid water) in lattice direction i at lattice position x (unit lu) and lattice time t (unit ts). In this chapter, the D2Q9 model, which is defined as a velocity model on a square, is used for two-dimensional simulation. The discrete velocity c_i in 9 directions is:

$$c_i = \begin{cases} 0 & i = 0 \\ \left(\cos\left[\frac{(i-1)\pi}{2}\right], \sin\left[\frac{(i-1)\pi}{2}\right]\right) & i = 1,2,3,4 \\ \sqrt{2}\left(\cos\left[\frac{(i-5)\pi}{2} + \frac{\pi}{4}\right], \sin\left[\frac{(i-5)\pi}{2} + \frac{\pi}{4}\right]\right) & i = 5,6,7,8 \end{cases} \tag{2}$$

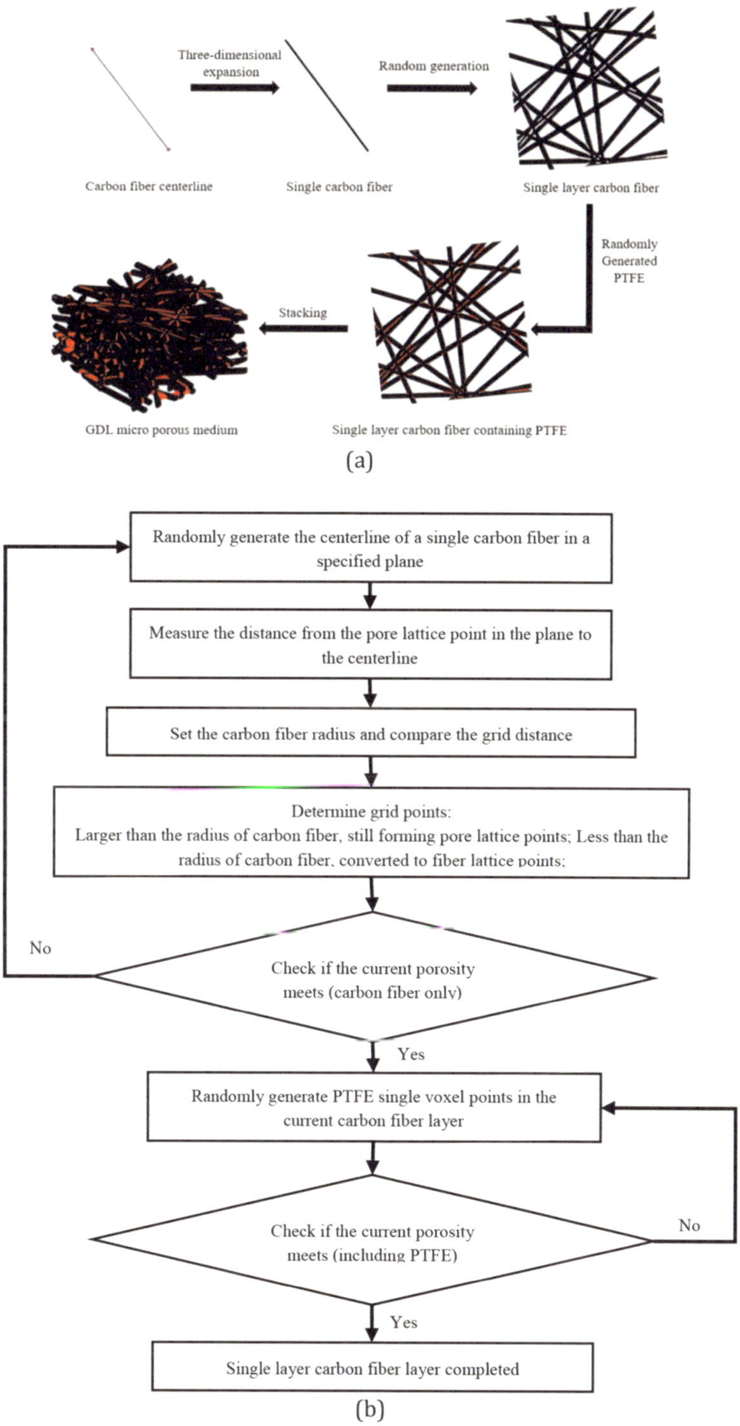

Figure 1.
(a) Distribution diagram and (b) flow diagram of GDL reconstruction.

For the DnQb model with single relaxation time (STR), the equilibrium distribution function $f_{\mathrm{i}}^{\mathrm{eq}}$ can be expressed as:

$$f_{\mathrm{i}}^{\mathrm{eq}} = \omega_{\mathrm{i}}\rho\left[1 + \frac{1}{c^2}(c_{\mathrm{i}} \cdot u) + \frac{1}{2c^4}(c_{\mathrm{i}} \cdot u)^2 - \frac{1}{2c^2}u^2\right] \tag{3}$$

Where c_s is the lattice sound velocity ($c_s = \sqrt{RT}$), and for the D2Q9 model, the weight factor ω_{i} is

$$\omega_{\mathrm{i}} = \begin{cases} \dfrac{4}{9} & i = 0 \\ \dfrac{1}{9} & i = 1, 2, 3, 4 \\ \dfrac{1}{36} & i = 5, 6, 7, 8 \end{cases} \tag{4}$$

In lattice units, τ (lattice unit ts) is defined as the relaxation time, which is related to the kinematic viscosity ν ($\mathrm{lu}^2\mathrm{ts}^{-1}$):

$$\nu = c_s{}^2(\tau_\nu - 0.5)\Delta t \tag{5}$$

The macroscopic density ρ and macroscopic velocity u of fluid component k can be obtained by the following equation:

$$\rho^k = \sum f_i^{\,k} \tag{6}$$

$$\rho^k u^k = \sum f_i^{\,k} e_i + \tau^k F^k \tag{7}$$

$$u^{\mathrm{k}} = u' + \frac{\tau^{\mathrm{k}} F^{\mathrm{k}}}{\rho_{\mathrm{k}}} \tag{8}$$

For multi-component and multiphase flow problems, the forces acting on various fluid components k are classified as surface tension $F_{1\mathrm{k}}$ between fluid and fluid, adhesion F_{2k} between fluid and solid, and external force F_{3k}. F^{k} represents the total force acting on component k:

$$F^{\mathrm{k}} = F_{1\mathrm{k}} + F_{2\mathrm{k}} + F_{3\mathrm{k}} \tag{9}$$

Therefore, for the component k particles at position x subjected to the force of fluid particles from position x':

$$F_{1\mathrm{k}}(x) = -\psi_{\mathrm{k}}(\rho_{\mathrm{k}}(x))\sum_{\mathrm{x}'}\sum_{\bar{\mathrm{k}}} G_{\mathrm{k}\bar{\mathrm{k}}}(x, x')\psi_{\mathrm{k}}(\rho_{\mathrm{k}}(x))(x' - x) \tag{10}$$

Where $\psi(\rho_{\mathrm{k}}(x))$ is the pseudo-potential function (also representing the effective density), defined as:

$$\psi(\rho_{\mathrm{k}}(x)) = \rho_0[1 - \exp(-\rho_{\mathrm{k}})\rho_0] \tag{11}$$

Where $G_{\mathrm{k}\bar{\mathrm{k}}}$ represents the strength of the interaction between the fluid and the fluid, considering only the forces of the nearest neighboring particles. By controlling the value of $G_{k\bar{k}}$, the insolubility and surface tension of the two-phase fluid are controlled.

When the fluid k particle at position x is transported to position x', it is subjected to a force from adjacent solid particles:

$$F_{2k}(x) = -\rho_k(x)\sum_{x'} G_{ks}(x, x')n_s(x')(x' - x) \tag{12}$$

Where G_{ks} represents the intensity of fluid-solid interaction between fluid particles and solid surfaces, and the different wetting characteristics between fluid and solid surfaces can be achieved by adjusting the value of G_{ks}. The n_s indicator function is used to distinguish each phase. When $n_s = 1$, it represents the solid phase, and when $n_s = 0$, it represents the liquid phase.

In this section, the SC pseudo-potential multiphase LB model is proposed to study the gas-liquid two-phase transport process in porous media and the influence of various structural design parameters on the dynamic transport process of liquid water.

3.2 LB method verification

Two numerical experiments were conducted in this section to validate the SC pseudo-potential multiphase LB method mentioned in Section 3.1 to simulate the physical phenomena of two-phase fluid flow in porous structures: one was a bubble experiment to obtain the gas-liquid interaction force $G_{k\bar{k}}$, and the other was a static contact angle experiment to determine the fluid-solid interaction force G_{ks}, as shown in Eqs. (10) and (12).

3.2.1 Bubble test

Surface tension is caused by the adhesion of liquid molecules at the liquid gas interface. It plays a key role in determining the flow of two-phase fluid in porous media, and is closely related to capillary action and wetting phenomenon. For liquid droplets suspended in gas, according to the well known Laplace law, the pressure difference at the liquid-gas interface is related to the radius of the bubble:

$$\Delta P = \frac{\sigma}{R} \tag{13}$$

Where σ is the surface tension, and ΔP is the pressure difference between the inside and outside of the bubble, expressed in lm lu^{-1} ts^{-2}.

A circular bubble with a radius of 20 lattices is initially located at the center of the 100×100 lattice domain, as shown in **Figure 2**. All outer sides are considered as periodic boundary conditions. Inside the droplet, the initial densities of air and water are set to 2.00 and 1.00×10^{-5}, respectively. Outside the droplet, the initial densities of air and water are set to 2.00 and 0.00, respectively. The difference between the maximum and minimum density of a gas is a function of $G_{k\bar{k}}$. **Figure 3** shows that when the value of $G_{k\bar{k}}$ reaches above 0.07, the density difference appears rapidly, reflecting the beginning of phase separation. Once $G_{k\bar{k}}$ exceeds 0.12, the density contrast will increase to the maximum value of 2.00. Therefore, $G_{k\bar{k}} = 0.12$ is used in the simulation.

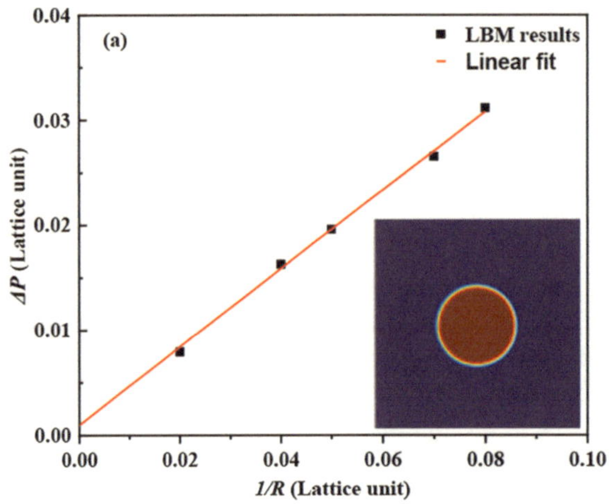

Figure 2.
Relationship between the internal and external pressure difference of the droplet and the reciprocal of the droplet radius.

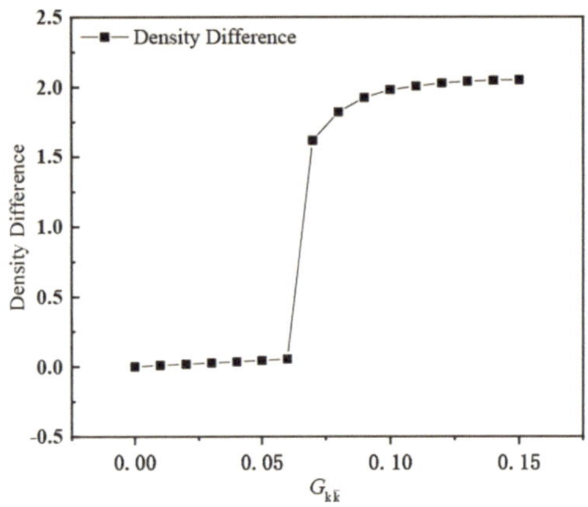

Figure 3.
Variation in the difference between the maximum and minimum densities of the gas with different values of $G_{k\bar{k}}$.

The pressure difference (ΔP) is at the liquid-gas interface (expressed in lattice units) and the bubble radius (R) at the final equilibrium state (expressed in lattice units). As shown in **Figure 2**, the results well comply with Laplace's law.

3.2.2 Static contact angle test

The wettability of solid surfaces can be reflected by the contact angle, as described in Section 3.1. By adjusting the value of G_{ks}, different wettability between fluid and solid surfaces can be achieved. In this simulation, semicircular droplets are placed on a horizontal solid plate to simulate the evolution of liquid shape on the solid wall at different G_{ks} values. The left and right sides of the calculation domain are considered as Periodic boundary conditions, while the upper and lower sides are set as no-slip boundary conditions. Similar to bubble testing, without any external force, the initial density of the fluid inside and outside the droplet is set to 2.00 and 1.0×10^{-5}, respectively. After reaching a stable state, the shape of the droplet and the angle of the

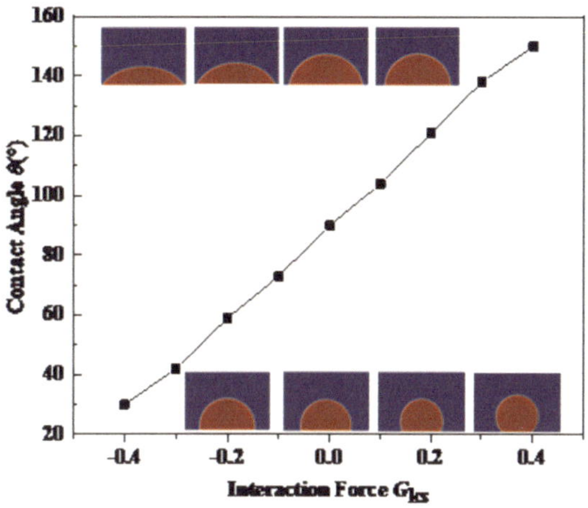

Figure 4.
Static contact angle as a function of G_{ks}.

liquid-solid interface were recorded using the method proposed in the reference literature to obtain the contact angle. In **Figure 4**, an approximate linear relationship between the predicted contact angle and G_{ks} can be observed.

4. Results and analysis

4.1 Influence of surface characteristics of porous structures

The wettability of solid phase in porous structure is an important characteristic that affects its internal capillary pressure. In order to explore the influence of changes in the internal surface characteristics of porous media structure on gas-liquid two-phase transport, the wettability of porous structure is changed by changing the content of PTFE in GDL.

To investigate the effect of solid phase wettability on the gas-liquid two-phase transport inside GDL, this section first investigated the dynamic transport behavior of liquid water in traditional GDL with five different PTFE contents (0 wt., 5 wt., 10 wt., 15 wt., and 20 wt.%). The porosity of GDL without PTFE treatment (0 wt.%) was set to 60%, while the overall porosity of GDL decreased after PTFE solution treatment. The PTFE content and structural design parameters in GDL are shown in **Table 1**.

Case Number	PTFE content (wt.%)	Porosity (%)
Case 1	0 wt.%	60.00%
Case 2	5 wt.%	56.80%
Case 3	10 wt.%	53.60%
Case 4	15 wt.%	50.40%
Case 5	20 wt.%	47.20%

Table 1.
Conventional GDL with different PTFE contents.

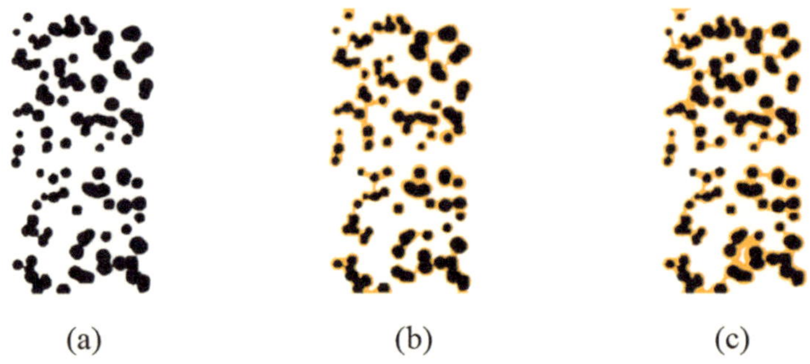

Figure 5.
Three representative reconstructed of the two-dimensional microstructure of GDLs with PTFE contents of (a) 0 wt. %, (b) 10 wt.%, and (c) 20 wt.%.

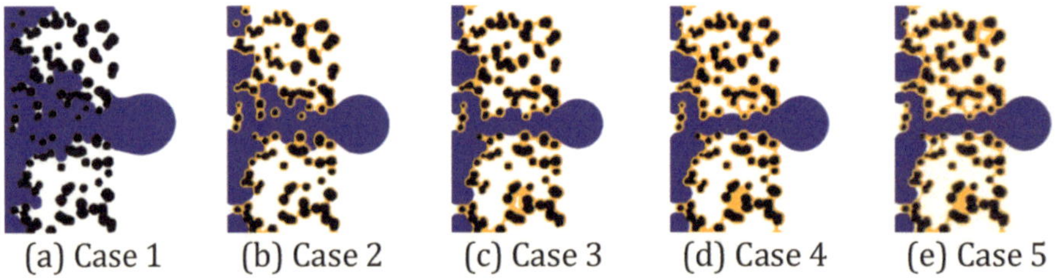

Figure 6.
Profiles of liquid water invading conventional GDLs with different PTFE contents in the steady state.

Figure 5 shows three representative reconstructed two-dimensional microstructures of GDL with PTFE content of 0 wt., 10 wt., and 20 wt.%.

This section discusses the dynamic behavior of two-phase fluid in five GDL porous structures with uniformly distributed PTFE content, and the structural design is shown in **Table 1**. **Figure 6** shows a cross-sectional view of liquid water invading GDL with different PTFE contents in steady-state. In all cases shown in **Figure 6**, capillary fingering flow can be observed in liquid water, and capillary pressure is the main driving force for liquid transport in porous structures. Liquid water tends to invade pores with minimal capillary resistance, such as hydrophilic channels with larger pores. After breaking through GDL, liquid water will form a fixed liquid water flow channel, and subsequent liquid water will first break through this channel. From **Figure 6(a)** and **(b)**, it can be seen that in untreated PTFE and GDL with low PTFE content, many large pores, even hydrophilic small pores, are occupied by liquid water. This can be attributed to the low capillary resistance caused by the hydrophilicity or low hydrophobicity of the solid phase, allowing liquid water to invade these pores. However, increasing the content of PTFE can lead to the need for liquid water to overcome greater capillary resistance, leading to the need for liquid water to break through larger pores rather than relatively smaller ones, which can effectively improve the transport of liquid water, as shown in **Figure 6(b)**–**(d)**).

From **Figure 7(a)**, it can also be seen that the higher the PTFE content, the lower the liquid water saturation. As shown in **Figure 7(b)**, as the PTFE content in GDL increases, the gas-liquid steady-state time decreases, indicating that the higher the hydrophobicity, the more favorable the drainage of GDL. However, the excessively hydrophobic porous structure significantly increases the entry pressure of liquid water, which may make the inlet side of GDL more susceptible to water flooding. It

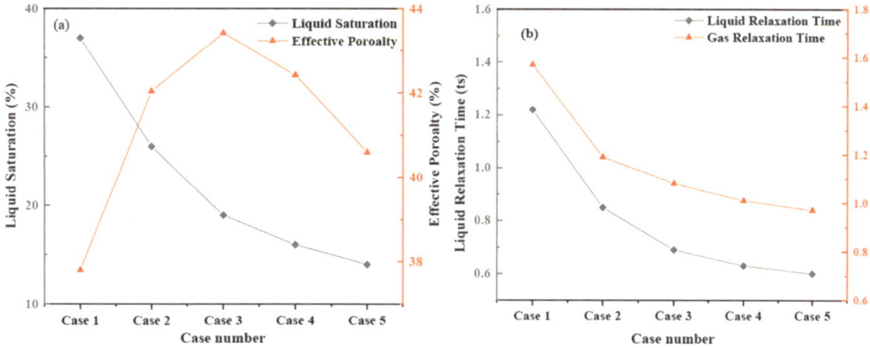

Figure 7.
(a) Curves of effective porosity and liquid saturation and (b) steady-state time curves of gas and liquid for conventional GDLs with different PTFE contents.

can be seen that evaluating the overall drainage performance of fuel cells solely based on liquid saturation is not enough. The main purpose of drainage is to enhance the gas mass diffusion of fuel cells. Therefore, effective porosity closely related to gas mass diffusion is used as another parameter for performance evaluation. Therefore, this study uses effective porosity as an additional criterion for evaluating GDL under different PTFE contents. As the PTFE content increases, the porosity of GDL decreases, but the effective porosity of GDL shows a trend of first increasing and then decreasing, as shown in **Figure 7(a)**. When the PTFE content is 10 wt.%, the effective porosity of GDL is the highest, reaching 43.42%. Compared to GDL without PTFE and with a PTFE content of 20 wt.%, it increases by 14.80 and 6.90%, respectively.

Therefore, for the porous media structure with capillary pressure as the main driving force, the solid phase is too hydrophilic, which will lead to the retention of liquid water in the pores, affecting the transport and discharge of liquid, while the excessively hydrophobic will inhibit the invasion of liquid phase, and the pores are mostly occupied by gas.

4.2 Influence of pore distribution in porous structures

The distribution of pore and fixed phase wettability has a significant impact on its multiphase transport. This section will construct a porous microstructure with pore gradient distribution and discuss the dynamic changes in its internal two-phase transport.

In this chapter, a two-dimensional profile is obtained from the three-dimensional GDL porous media structure constructed using the random algorithm in the second section. The above random algorithm was used to design and reconstruct GDLs with PTFE gradient distribution, including PTFE bi-gradient and tri-gradient GDLs. The inlet and outlet regions of PTFE bi-gradient GDL have different PTFE contents, while the inlet, middle, and outlet regions of PTFE tri-gradient GDL have different PTFE contents. **Figure 8** shows the representative reconstructed two-dimensional microstructure of bi-gradient and tri-gradient GDL with a total PTFE content of 10 wt.%. For bi-gradient GDL, as shown in **Figure 8(a)**, the corresponding PTFE content in the inlet and outlet regions of GDL is 14 wt. and 6 wt.%. Similarly, **Figure 8(b)** shows the three-dimensional microstructure of GDL with different PTFE contents in the inlet, middle, and outlet regions.

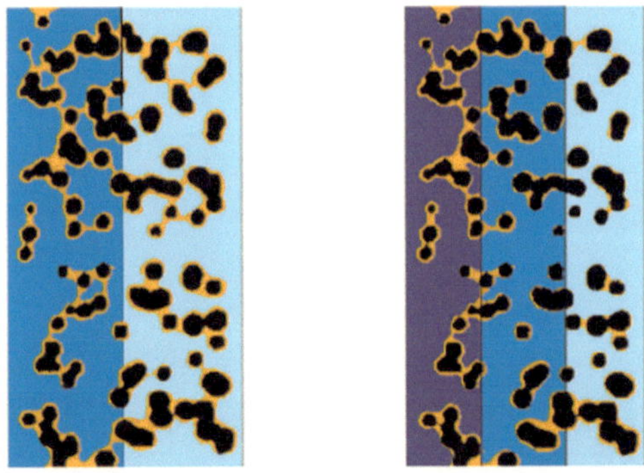

Figure 8.
Representative reconstructed microstructure of bi-gradient and tri-gradient PTFE GDLs with a total PTFE content of 10 wt.%: (a) inlet regions: 14 wt.%, outlet regions: 6 wt.%, and (b) inlet regions: 16 wt.%, middle regions: 10 wt.%, outlet regions: 4 wt.%.

Case number	PTFE content (wt.%)	Region width (μm)	Porosity
Case 6	in: 4, out:16	in: 50, out: 50	53.60%
Case 7	in: 6, out: 14	in: 50, out: 50	53.60%
Case 8	in: 8, out: 12	in: 50, out: 50	53.60%
Case 9	in: 12, out: 8	in: 50, out: 50	53.60%
Case 10	in: 14, out: 6	in: 50, out: 50	53.60%
Case 11	in: 16, out: 4	in: 50, out: 50	53.60%

Notes: "in" and "out" represent the inlet region and out region of the gradient PTFE GDL.

Table 2.
Bi-gradient PTFE GDL with the total PTFE content of 10 wt.%.

This section proposes six GDL structural designs for bi-gradient PTFE with a total PTFE content of 10 wt.% and constructs a GDL porous structure with gradient distribution of wetting characteristics. The design scheme and parameters of the bi-gradient GDL structure with PTFE content are shown in **Table 2**.

On this basis, the performance comparison simulation of PTFE bi-gradient GDL and traditional GDL (Case 3) was conducted using SC multiphase pseudo-potential LBM. **Figure 9** shows the profile of PTFE content bi-gradient GDL (Case 6–11) and traditional GDL (Case 3) with PTFE content of 10 wt.% under stable state when liquid water invades two equally sized segmentation regions. It can be observed that liquid water invades and exits the GDL structure under capillary pressure. However, in **Figure 9(a)–(c)**, the PTFE content in the treated bi-gradient GDL inlet area is relatively low, and the capillary resistance that liquid water needs to overcome when invading the pores of the GDL inlet area is low. The low hydrophobicity leads to relatively high adhesion, making liquid water occupy these pores and unable to be discharged. Therefore, compared to traditional GDL (Case 3), more liquid water occupies the pores of the inlet area. The liquid saturation in the inlet area is higher than that of traditional GDL structures. When liquid water begins to enter the pores of the outlet area with high PTFE content, the capillary resistance suddenly increases. In

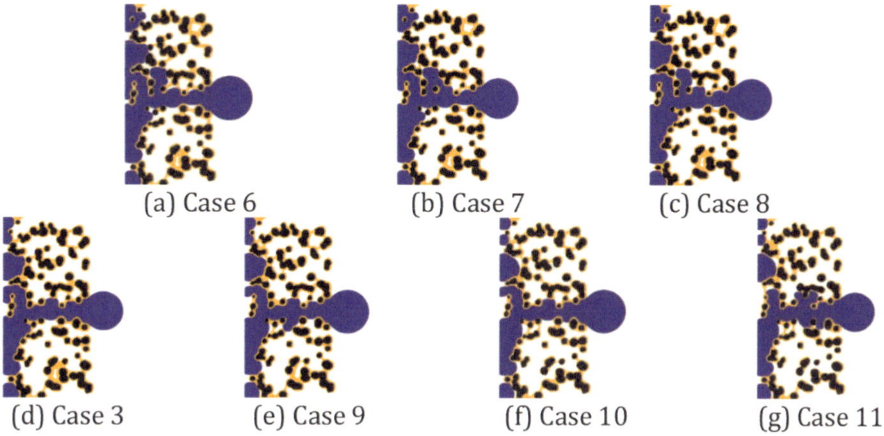

Figure 9.
Profiles of liquid water invading the bi-gradient PTFE GDLs with two regions of the same size and the conventional GDL with a PTFE content of 10 wt.% at steady state.

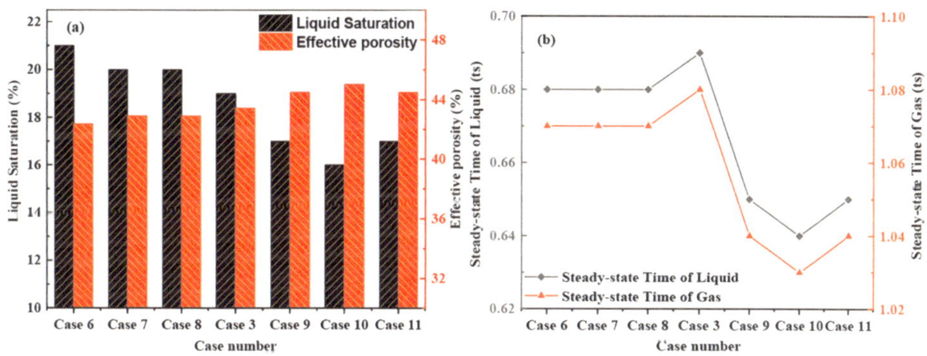

Figure 10.
(a) Curves of effective porosity and liquid water saturation and (b) steady-state time curves of gas and liquid for the bi-gradient GDL with two regions of the same size and the conventional GDL with a PTFE content of 10 wt.%.

this case, the outlet area with high PTFE content acts as a capillary barrier, hindering the breakthrough of liquid water, which in turn increases the residual water content in the pores of the inlet area. Therefore, cases 6 and 7 have higher water saturation and lower effective porosity, and the breakthrough time of liquid water from GDL is longer. As shown in **Figure 10(a)** and **(b)**, the liquid saturation in cases 6–8 is relatively high, and more liquid water occupies and invades the pores, thus hindering the gas mass diffusion ability, reducing the effective porosity. The structure is unfavorable to the drainage performance and gas diffusion performance of porous media.

In **Figure 9(e)**–**(g)**, compared to the traditional GDL structure design, the treated GDL has a relatively high PTFE content in the inlet area and a relatively high hydrophobic property. The capillary barrier formed in the inlet area has hindered the breakthrough and occupation of liquid water, resulting in less residual liquid water content in the pores of the inlet area. At the same time, the pores with lower hydrophobicity in the outlet area reduce the capillary resistance overcome by liquid water breakthrough, which is more conducive to the breakthrough and discharge of liquid water. As shown in **Figure 9(a)** and **(b)**, compared to the performance parameters of traditional GDL (Case 3), PTFE with a higher content in the inlet area has a lower design liquid saturation and higher effective porosity. At the same time, the lower liquid equilibrium time in **Figure 10(b)** indicates that the time for liquid breakthrough and discharge of GDL is lower, and the drainage performance is improved. Therefore, a higher distribution of PTFE content in the inlet section is beneficial for the liquid transport of GDL. However, as shown in Case 11 in **Figure 10(a)**, the high PTFE content in the inlet area means that the PTFE content in the outlet area is too low. In this case, its extremely low hydrophobicity leads to more liquid water occupying the pores of the outlet area, which is not conducive to the discharge of liquid water and the transportation efficiency of the gas. In addition, a high PTFE content in the inlet area can also lead to a high inlet pressure when liquid water enters GDL, hindering the liquid water from entering GDL from the inlet side and discharging into the fuel cell, thereby making the inlet side more susceptible to water flooding.

Similarly, this section discusses three GDL structures with tri-gradients of PTFE content, with higher PTFE content distributed near the inlet area. The PTFE content in the middle region remains at 10 wt.%. Therefore, when the average PTFE content is 10 wt.%, the PTFE content in the export area is determined by the PTFE content in the import area. The tri-gradient structure design is shown in **Table 3**.

Figure 11 shows the GDL structure with three gradients of PTFE content when liquid water invades three regions of the same size during steady state, as well as a cross-sectional view of the traditional GDL of Case 3 and the GDL with bi-gradients of PTFE content of Case 10. According to the discussion and explanation in Section 4.2, a high PTFE content (14 wt.%) in the inlet area is beneficial for the drainage of GDL, while a low or high PTFE content in the inlet area is not conducive to the liquid transport of GDL. Similar conclusions can be applied to the GDL of PTFE with three gradients, as shown in cases 12 and 13 in **Figure 11**. However, for the PTFE content three gradient GDL, a reasonable PTFE content in the inlet area (14 wt.%), compared

Case number	PTFE content (wt.%)	Region width (μm)	Porosity
Case 12	in: 12, mid: 10, out: 8	in: 33.3, mid: 33.3, out 33.3	53.60%
Case 13	in: 14, mid: 10, out: 6	in: 33.3, mid: 33.3, out 33.3	53.60%
Case 14	in: 16, mid: 10, out: 4	in: 33.3, mid: 33.3, out 33.3	53.60%

Notes: "in," "mid," and "out" represent the inlet region, middle region, and outer region of the gradient PTFE GDL, respectively.

Table 3.
Tri-gradient PTFE GDL with a total PTFE content of 10 wt.%.

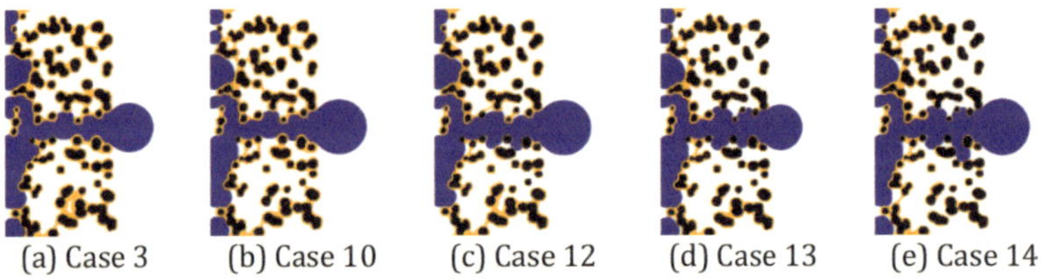

Figure 11.
Profile of liquid water intrusion into GDL with PTFE tri-gradient under stable-state time (conventional GDL in case 3 and PTFE bi-gradient GDL in Case10).

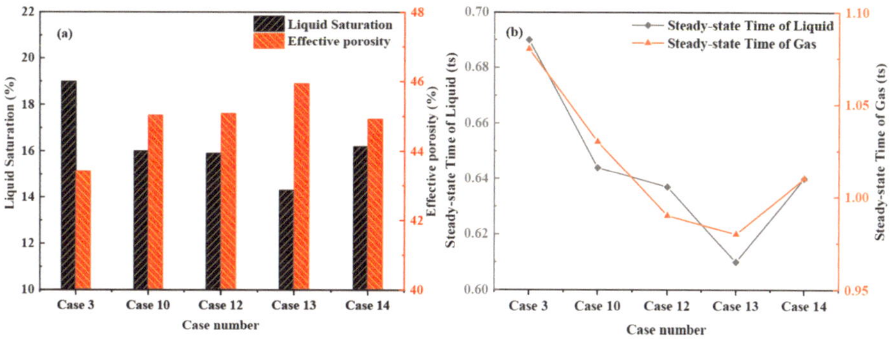

Figure 12.
Steady-state time of (a) curves of effective porosity and liquid saturation degree and (b) gas and liquid for the different tri-gradient GDLs with three regions of the same sizes, the conventional GDL of case 3 and the bi-gradient PTFE GDL of case 10 at steady state.

to the traditional GDL in Case 3 and the optimal PTFE bi-gradient GDL in Case 10, resulted in an increase in effective porosity of 5.80 and 2.00%, respectively, as shown in **Figure 12**. Due to the presence of an intermediate region with a PTFE content of 10 wt.%, the size of the outlet region with relatively low PTFE content is reduced, resulting in less residual water in the pores of the outlet region. In addition, the middle region of the PTFE tri-gradient GDL reduces the PTFE content gradient difference between the inlet region (14 wt.%) and the outlet region (6 wt.%). In this case, the PTFE tri-gradient GDL generates relatively smooth capillary resistance, which may be beneficial for the drainage of liquid water. As shown in **Figure 12(b)**, the steady-state time curves of gases and liquids can also support this point.

Therefore, the gradient distribution of wettability will make the capillary pressure distribution in local regions uneven, or even form a capillary barrier, thus changing the two-phase transport state. It is of great significance to reasonably distribute the wettability of solid phase for improving the internal transport of porous media.

4.3 Effects of porous structure aging

The porous structure will cause structural damage and changes in the surface characteristics of the solid phase due to corrosion, shedding, and other conditions, thus affecting its internal multiphase flow state. In order to obtain the influence mechanism of the aging of porous media structure on two-phase transport, this section will conduct aging treatment and study on GDL structure.

To study the influence mechanism of porous media aging on liquid water transport behavior, the GDL of carbon fiber and PTFE degraded at the same time with different content of PTFE was reconstructed. Assuming that the final degradation amounts of carbon fiber and PTFE after 20 aging cycles in GDL are 30 and 15%, respectively, four different aging levels of GDL were reconstructed using exponential rate random degradation for 5, 10, 15, and 20 aging cycles, simulating liquid water transport in GDL under different aging levels. **Figure 13(a)–(c)** shows the reconstruction results of fresh GDL and aged GDL subjected to 10 cycles of aging. **Figure 13(f)** shows a large image of the local GDL component degradation. Compared to fresh GDL, aged GDL exhibits corrosion of carbon fibers and detachment of PTFE. The structural design scheme is shown in **Table 4**.

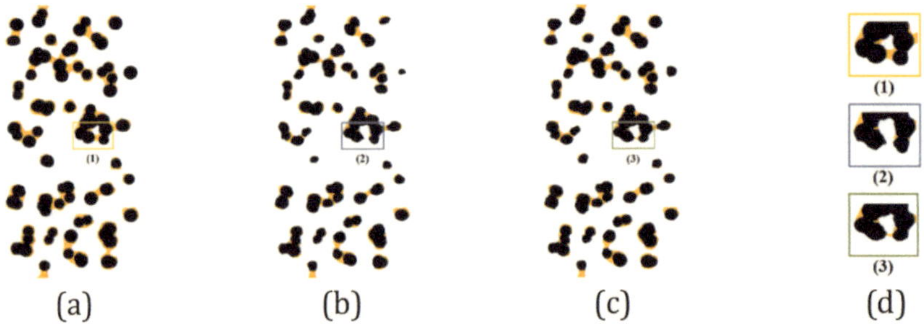

Figure 13.
(a)–(c) fresh GDL and GDL with different aging processes, and (f) enlarged view of local aging.

PTFE Content (wt.%)	Case Number	Carbon fiber degradation amount (%)	PTFE degradation amount (%)
5 wt.%	Case 15	0%	0.00%
	Case 16 (5-times)	2%	1.61%
	Case 17 (10-times)	6%	4.20%
	Case 18 (15-times)	14%	8.37%
	Case 19 (20-times)	30%	15.00%
10 wt.%	Case 20	0%	0.00%
	Case 21 (5-times)	2%	1.61%
	Case 22 (10-times)	6%	4.20%
	Case 23 (15-times)	14%	8.37%
	Case 24 (20-times)	30%	15.00%
15 wt.%	Case 25	0%	0.00%
	Case 26 (5-times)	2%	1.61%
	Case 27 (10-times)	6%	4.20%
	Case 28 (15-times)	14%	8.37%
	Case 29 (20-times)	30%	15.00%

Table 4.
Fresh GDL and aged GDL with different initial PTFE contents under exponential rate degradation.

Figure 14 shows the changes in the transport behavior of liquid water in fresh GDL and aged GDL as the aging degree increases when the initial PTFE content is 5 wt., 10 wt., and 15 wt.%. The increase in initial PTFE content can occupy some pores, resulting in a decrease in total porosity and an increase in capillary pressure. In addition, more carbon fibers are covered with PTFE, leading to an increase in the hydrophobic pathway of GDL and a further increase in capillary pressure. Therefore, GDL with a high initial PTFE content produces lower liquid water saturation than GDL with a low initial PTFE content. As GDL ages, higher PTFE content also leads to lower liquid water saturation in GDL compared to GDL structures with lower PTFE content, as shown in Cases 19, 24, and 29 in **Figure 14**.

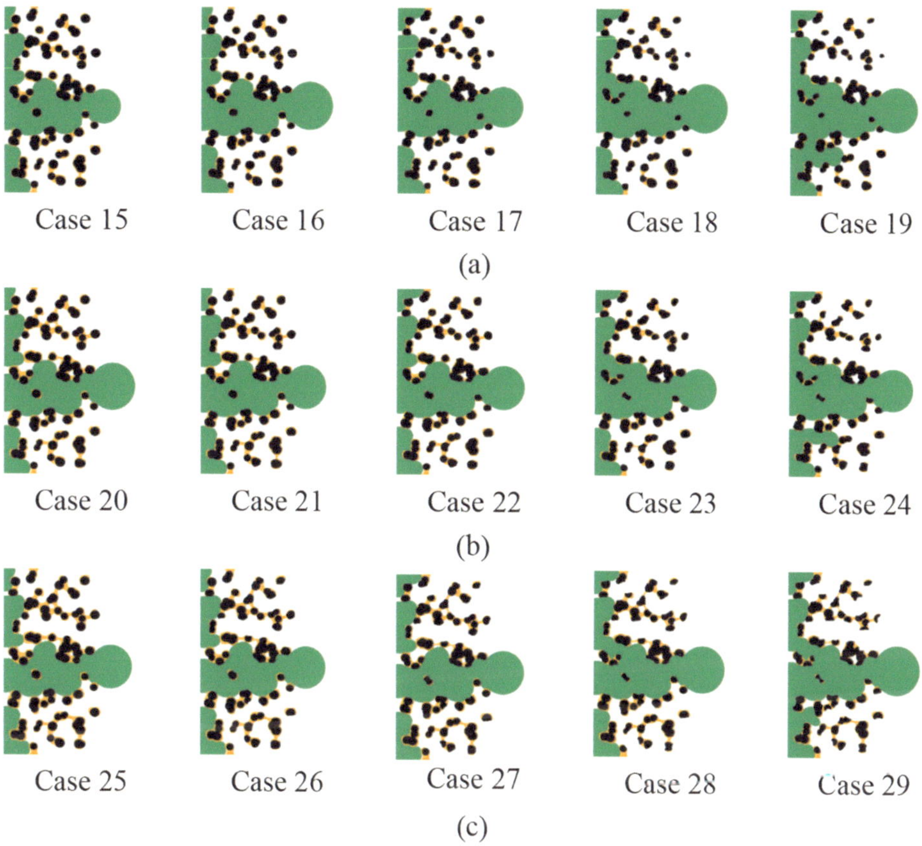

Figure 14.
Profile of liquid water invading a fresh GDL and GDLs aged with initial PTFE contents of (a) 5 wt.%, (b) 10 wt.%, and (c) 15 wt.%.

From **Figure 15(a)**, it can be seen that increasing the initial PTFE content can indeed reduce the liquid water saturation in GDL, but it will also occupy the overall porosity of GDL and reduce the effective porosity, thereby reducing the gas mass diffusion ability. In traditional GDL designs with different initial PTFE contents, the saturation of liquid water increases with increasing aging degree, while the corresponding effective porosity continues to decrease. However, during the entire aging process of GDL, GDL with an initial PTFE content of 10 wt.% showed the highest effective porosity, corresponding to the best overall performance. At the end of aging, the liquid saturation of GDL with a PTFE content of 10 wt.% increased by 10.20% compared to the initial GDL, but the effective porosity decreased by 4.20%. When the PTFE content is 15 wt.%, the liquid saturation increases by 9.1% and the effective porosity decreases by 4.30%. In addition, excessively hydrophobic GDL can lead to high inlet pressure of liquid water, resulting in liquid water generated by electrochemical reactions in CL remaining in CL and not being discharged from the fuel cell interior through GDL. From **Figure 15(b)**, it can be seen that GDL with an initial PTFE content of 10 wt.% has a higher gas-liquid steady-state time and better drainage performance.

The above results show that aging will destroy the structure of porous media, reduce the overall capillary pressure, and even change the hydrophilic and

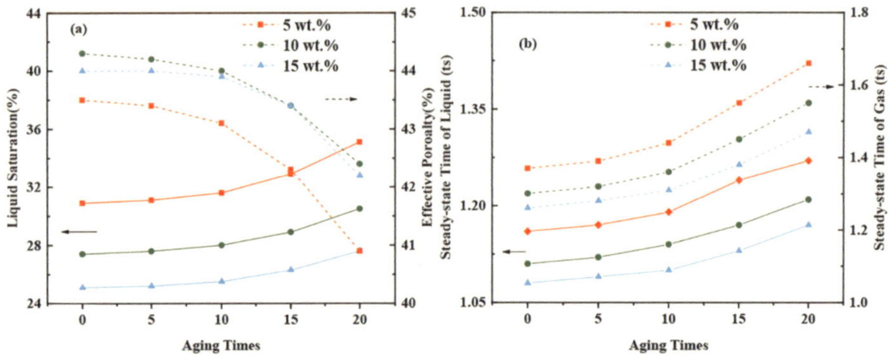

Figure 15.
Curves for (a) liquid saturation and effective porosity and (b) gas and liquid steady-state time for a fresh GDL and GDLs aged with initial PTFE contents of 5 wt.%, 10 wt.%, and 15 wt.%.

hydrophobic properties of local solid phases, making liquid water accumulate in local pores and unable to be discharged, seriously affecting the multiphase transport efficiency inside porous media. Therefore, improving durability is an important way to ensure the internal transport of porous media structures.

5. Conclusion

This chapter introduces the construction of microporous media structure through random algorithm and uses the SC pseudo potential LB method to study multiphase transport in porous media. Taking the GDL structure of PEM fuel cell as an example, the structural characteristics of porous media, such as solid phase wettability, pore distribution, and structural aging, are deeply discussed, and structural aging, were discussed in depth, and the following conclusions were obtained:

1. The higher the PTFE content in GDL porous media structure, the shorter the liquid water penetration time, but the effective porosity of GDL decreases. When the PTFE content reaches 10 wt.%, GDL has the maximum effective porosity.

2. For gradient GDL, a reasonably high PTFE content near the inlet will reduce the liquid water penetration time, and the effective porosity of bi-gradient (14 wt.%/ 6 wt.%) and tri-gradient (14 wt.%, 10 wt.%, and 6 wt.%) GDL will increase by 4.20 and 5.60%.

3. The initial PTFE content in GDL will affect the anti-aging performance of GDL. Higher PTFE content can ensure that GDL has better anti-aging performance, but too high initial PTFE content will also affect the gas mass diffusion ability.

This chapter obtained the intrinsic relationship between structural characteristic parameters and multiphase fluid flow is obtained, this lays a theoretical foundation for studying gas-liquid transport in microporous structures.

Acknowledgements

We thank the supports for this project provided by National Natural Science Foundation of China (No. 52176084) and National Key Research and Development Program of China (No. 2022YFE0207600).

Conflict of interest

The authors declare no conflict of interest.

Appendices and nomenclature

c_s	speed of sound in lattice, lu ts^{-1}
d	pore diameter, lu
e_i	velocity of particles in lattice direction i, lu ts^{-1}
F^k	total force acting on k component, lm lu ts^{-2}
F_{1k}	fluid/fluid interaction force acting on k component, lm lu ts^{-2}
F_{2k}	fluid/solid interaction force acting on k component, lm lu ts^{-2}
F_{3k}	other external forces acting on the k component, lm lu ts^{-2}
f_i^k	density distribution function of k component in lattice direction i
$f_i^{eq,k}$	density equilibrium distribution function of k component in lattice direction i
$G_{k\bar{k}}$	cohesion factor between components k and $\bar{k}$
G_{ks}	adhesion factor between component k and solid
$J(s)$	Leverett J function
P	pressure, lm lu^{-1} ts^{-2}
ΔP	pressure difference inside and outside the bubble, lm lu^{-1} ts^{-2}
R	droplet radius, lu
t	lattice time, ts
Δt	lattice time step, ts
u'	macroscopic velocity, lu ts^{-1}
u^{eq}	lattice velocity of component k, lu ts^{-1}
u^k	equilibrium velocity of component k, lu ts^{-1}
ω_i	weight factor in lattice direction i
x	lattice position, lu

Greek symbols

τ	relaxation time, ts
ν	kinematic viscosity, lu^2 ts^{-1}
ρ	density, lm lu^{-3}
ψ	effective density
σ	surface tension
θ	static contact angle
ε	porosity

Lattice unit

lm	lattice unit of mass
lu	lattice unit of length
ts	lattice time step

Author details

Yulin Wang[1,2]* and Haokai Xu[1]

1 Tianjin Key Lab of Refrigeration Technology, Tianjin University of Commerce, China

2 Ningbo Institute of Material Technology and Engineering, Chinese Academy of Sciences, Ningbo, Zhejiang Province, China

*Address all correspondence to: wangylfcs@gmail.com

References

[1] Pesavento F, Schrefler BA, Sciumè G. Multiphase flow in deforming porous media: A review. Archives of Computational Methods in Engineering. 2017;**24**:423-448

[2] Kasaeian A, Daneshazarian R, Mahian O, Kolsi L, Chamkha AJ, Wongwises S, et al. Nanofluid flow and heat transfer in porous media: A review of the latest developments. International Journal of Heat and Mass Transfer. 2017;**107**:778-791

[3] Koohbor B, Colombano S, Harrouet T, Deparis J, Lion F, Davarzani D, et al. The effects of water table fluctuation on LNAPL deposit in highly permeable porous media: A coupled numerical and experimental study. Journal of Contaminant Hydrology. 2023;**256**:104183

[4] Colombano S, Davarzani H, Van Hullebusch ED, Huguenot D, Guyonnet D, Deparis J, et al. Comparison of thermal and chemical enhanced recovery of DNAPL in saturated porous media: 2D tank pumping experiments and two-phase flow modelling. Science of the Total Environment. 2021;**760**:143958

[5] Luévano-Rivas OA, Quiroz-Ramirez JJ, Suarez-Toriello VA, Huerta-Rosas B, Sánchez-Ramirez E, Segovia-Hernández JG. Upscaling of mass and heat transport applied to reactive packing catalytic porous media. Chemical Engineering Science. 2023;**265**:118206

[6] Zhao J, Tu Z, Chan SH. Carbon corrosion mechanism and mitigation strategies in a proton exchange membrane fuel cell (PEMFC): A review. Journal of Power Sources. 2021;**488**:229434

[7] Koch T, Weishaupt K, Müller J, Weigand B, Helmig R. A (dual) network model for heat transfer in porous media: toward efficient model concepts for coupled systems from fuel cells to heat exchangers. Transport in Porous Media. 2021;**140**(1):107-141

[8] Zhang Y, Tao Y, Shao J. Application of porous materials for the flow field in polymer electrolyte membrane fuel cells. Journal of Power Sources. 2021;**492**:229664

[9] Ben-Noah I, Friedman SP, Berkowitz B. Dynamics of air flow in partially water-saturated porous media. Reviews of Geophysics. 2023;**61**:e2022RG000798

[10] Chen L, He A, Zhao J, Kang Q, Li Z, Carmeliet J, et al. Pore-scale modeling of complex transport phenomena in porous media. Progress in Energy and Combustion Science. 2022;**88**:100968

[11] Peighambardoust SJ, Rowshanzamir S, Amjadi M. Review of the proton exchange membranes for fuel cell applications. International Journal of Hydrogen Energy. 2010;**35**(17):9349-9384

[12] Ijaodola OS, El-Hassan Z, Ogungbemi E, Khatib FN, Wilberforce T, Thompson J, et al. Energy efficiency improvements by investigating the water flooding management on proton exchange membrane fuel cell (PEMFC). Energy. 2019;**179**:246-267

[13] Kandlikar SG, Garofalo ML, Lu Z. Water management in a PEMFC: Water

transport mechanism and material degradation in gas diffusion layers. Fuel Cells. 2011;**11**(6):814-823

[14] Omrani R, Shabani B. Gas diffusion layer modifications and treatments for improving the performance of proton exchange membrane fuel cells and electrolysers: A review. International Journal of Hydrogen Energy. 2017; **42**(47):28515-28536

[15] Chun JH, Park KT, Jo DH, Kim SG, Kim SH. Numerical modeling and experimental study of the influence of GDL properties on performance in a PEMFC. International Journal of Hydrogen Energy. 2011;**36**(2):1837-1845

[16] Su H, Sita C, Pasupathi S. The effect of gas diffusion layer PTFE content on the performance of high temperature proton exchange membrane fuel cell. International Journal of Electrochemical Science. 2016;**11**(4):2919-2926

[17] Yu J, Froning D, Reimer U, Lehnert W. Apparent contact angles of liquid water droplet breaking through a gas diffusion layer of polymer electrolyte membrane fuel cell. International Journal of Hydrogen Energy. 2018; **43**(12):6318-6330

[18] Wang XL, Qu ZG, Lai T, Ren GF, Wang WK. Enhancing water transport performance of gas diffusion layers through coupling manipulation of pore structure and hydrophobicity. Journal of Power Sources. 2022;**525**:231121

[19] Park J, Oh H, Ha T, Lee YI, Min K. A review of the gas diffusion layer in proton exchange membrane fuel cells: Durability and degradation. Applied Energy. 2015;**155**:866-880

[20] Mortazavi M, Tajiri K. Effect of the PTFE content in the gas diffusion layer on water transport in polymer electrolyte fuel cells (PEFCs). Journal of Power Sources. 2014;**245**:236-244

[21] Tsai JC, Lin CK. Effect of PTFE content in gas diffusion layer based on Nafion®/PTFE membrane for low humidity proton exchange membrane fuel cell. Journal of the Taiwan Institute of Chemical Engineers. 2011;**42**(6): 945-951

[22] Hinebaugh J, Lee J, Bazylak A. Visualizing liquid water evolution in a PEM fuel cell using synchrotron X-ray radiography. Journal of the Electrochemical Society. 2012;**159**(12): F826

[23] Cooper NJ, Santamaria AD, Becton MK, Becton MK. Neutron radiography measurements of in-situ PEMFC liquid water saturation in 2D & 3D morphology gas diffusion layers. International Journal of Hydrogen Energy. 2017;**42**(25):16269-16278

[24] Meyer Q, Ashton S, Boillat P, Cochet M, Engebretsen E, Finegan DP, et al. Effect of gas diffusion layer properties on water distribution across air-cooled, open-cathode polymer electrolyte fuel cells: A combined *ex-situ* X-ray tomography and in-operando neutron imaging study. Electrochimica Acta. 2016;**211**:478-487

[25] Yu S, Li X, Liu S, Hao J, Shao Z, Yi B. Study on hydrophobicity loss of the gas diffusion layer in PEMFCs by electrochemical oxidation. RSC Advances. 2014;**4**(8):3852-3856

[26] Allen MB, Behie GA, Trangenstein JA. Multiphase Flow in Porous Media: Mechanics, Mathematics, and Numerics. Berlin: Springer-Verlag, Springer Science & Business Media; 2013

[27] Nield DA, Bejan A, Nield DA. Mechanics of fluid flow through a porous

medium. In: Convection in Porous Media. Cham: Springer; 2017. pp. 1-35

[28] Yang G, Weigand B. Investigation of the Klinkenberg effect in a micro/nanoporous medium by direct simulation Monte Carlo method. Physical Review Fluids. 2018;**3**(4): 044201

[29] Rabbani A, Babaei M, Javadpour F. A triple pore network model (T-PNM) for gas flow simulation in fractured, micro-porous and meso-porous media. Transport in Porous Media. 2020;**132**: 707-740

[30] Liu H, Kang Q, Leonardi CR, Schmieschek S, Narváez A, Jones BD, et al. Multiphase lattice Boltzmann simulations for porous media applications: A review. Computational Geosciences. 2016;**20**:777-805

[31] He YL, Liu Q, Li Q, Tao W. Lattice Boltzmann methods for single-phase and solid-liquid phase-change heat transfer in porous media: A review. International Journal of Heat and Mass Transfer. 2019; **129**:160-197

[32] Eshghinejadfard A, Daróczy L, Janiga G, Thévenin D. Calculation of the permeability in porous media using the lattice Boltzmann method. International Journal of Heat and Fluid Flow. 2016;**62**: 93-103

[33] Ira Y, Bakhshan Y, Khorshidimalahmadi J. Effect of wettability heterogeneity and compression on liquid water transport in gas diffusion layer coated with microporous layer of PEMFC. International Journal of Hydrogen Energy. 2021;**46**(33):17397-17413

[34] Han B, Ni M, Meng H. Three-dimensional lattice Boltzmann simulation of liquid water transport in porous layer of PEMFC. Entropy. 2015; **18**(1):17

[35] Jithin M, Siddharth S, Das MK, De A. Simulation of coupled heat and mass transport with reaction in PEM fuel cell cathode using lattice Boltzmann method. Thermal Science and Engineering Progress. 2017;**4**:85-96

[36] Zhou C, Guo L, Chen L, Tian X, He T, Yang Q. Pore-scale modeling of air–water two phase flow and oxygen transport in gas diffusion layer of proton exchange membrane fuel cell. Energies. 2021;**14**(13):3812

Chapter 4

Design Strategy of Metal Foam Flow Field and Its Application in Proton Exchange Membrane Fuel Cell

Zhongmin Wan, Yun Sun and Chen Yang

Abstract

Metal foam as flow field for proton exchange membrane fuel cell (PEMFC) has demonstrated substantial promise for enhancing the distribution uniformities of reactants and temperature. In different types of PEMFC, however, the influence of metal foam flow field on cell performance is diverse. For liquid-cooled PEMFC, employing metal foam flow field encounters challenges in water removal. For air-cooled PEMFC, employing metal foam can help retain a certain amount of liquid water to avoid membrane dehydration. In an effort to further enhance the cell performance and practical application potential of metal foam in different types of PEMFC, the design strategy for metal foam flow field in PEMFCs is studied through this chapter. Experimental results revealed that for liquid-cooled PEMFC, the cathode side employing metal foam flow field raises the potential for water flooding and instability of operational voltage, which can be addressed through reasonably designing the structural characteristics of metal foam. Furthermore, the thermal management capability of air-cooled PEMFC can be boosted through adopting metal foam flow field, owing to the synergic improvement of forced convective heat transfer of reactant gas and electrochemical performance, which is the main factor of the thermal management improvement at high current.

Keywords: PEMFC, metal foam flow field, design strategy, water management, thermal management

1. Introduction

Due to the global desire for more efficient and environmentally friendly energy usage, proton exchange membrane fuel cell (PEMFC), which is being recognized as a potential substitute for internal combustion engines, has gained increasing interest of scholars owing to its remarkable attributes such as excellent efficiency, low operating temperature, quickly starting speed, and environmentally clean operation [1]. The electrochemical mechanism and components of different types of PEMFC are

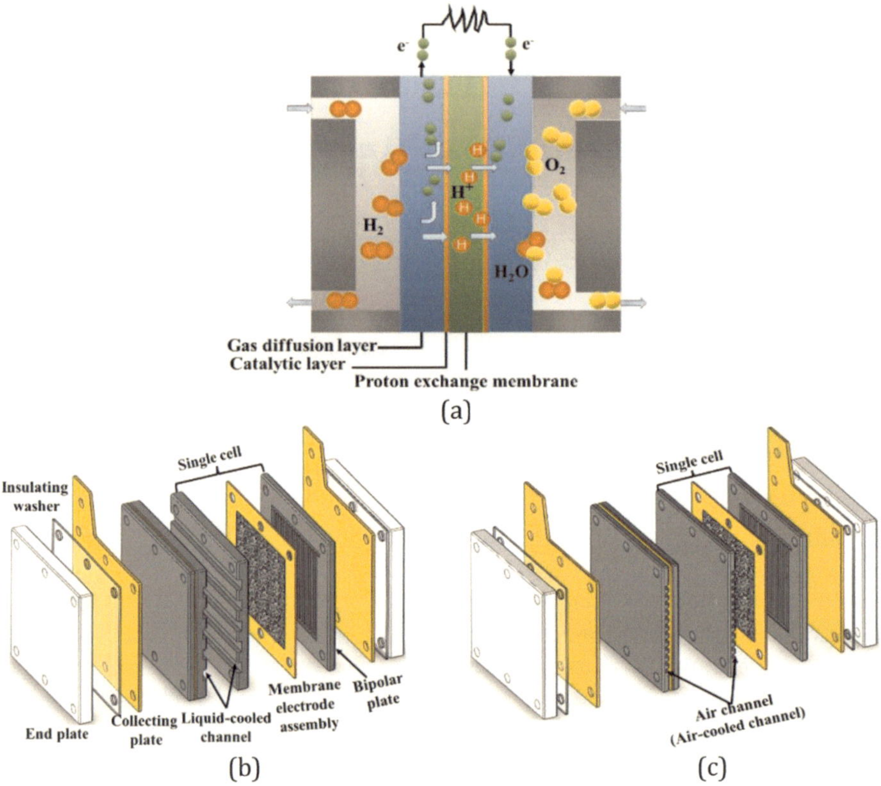

Figure 1.
The electrochemical mechanism and components of different types of PEMFC. (a) Schematics of the electrochemical mechanism of PEMFC. (b) Liquid-cooled PEMFC. (c) Air-cooled PEMFC.

depicted in **Figure 1**. In the past few years, PEMFC has experienced significant advancements and has found diverse applications in areas such as spacecraft, emergency supply, and transportation [2]. Enhancing the performance of PEMFC is crucial to expand its application areas.

Bipolar plate (BP), an essential constituent of PEMFC, plays a pivotal role in providing structural support and anchoring the membrane electrode assembly (MEA). Apart from these functions, flow field in BP performs additional vital functions, like distributing the reactants and products, collecting current and discharging, and so forth. The flow field design and optimization in BP play a crucial role in determining the performance of PEMFCs. Consequently, numerous scholars [3–5] have dedicated their efforts to designing and optimizing flow field for boosting the output performance of PEMFCs. Yan et al. [6] presented an innovative 3D serpentine flow channel that equips with a gradient wave structure. According to numerical and experimental results, in comparison with traditional serpentine flow channel, this innovative 3D serpentine flow channel exhibits the potential for enhancing the diffusion of the reactants to the catalytic layer (CL). Moreover, this innovative flow channel has the capability to augment the flow velocity, thus effectively removing excess liquid water. However, conventional flow field design in BPs frequently results in inhomogeneous gas distribution on the MEA particularly beneath the ribs. This, in result, leads to inhomogeneous electrochemical reaction and localized areas of elevated temperature at the MEA.

The phenomenon of localized areas of elevated temperature is detrimental to the durability of the MEA. In recent times, there has been considerable interest in utilizing metal foam as flow distributor for PEMFC owing to its exceptional distribution properties.

Metal foam exhibits remarkable electrical and thermal conductive properties, in addition to possessing high porosity as well as a substantial specific surface area [7, 8]. The reactant distribution uniformity, thermal management, and performance of PEMFC can be effectively boosted by the adoption of metal foam flow field. The study conducted by Afshari et al. [9] involved a numerical comparison of the performances of PEMFCs equipped with a partially restricted flow field as well as metal foam flow field. The findings demonstrated that the distribution uniformities of current density and oxygen concentration at the CL can be boosted by the implementation of metal foam flow field. In addition, based on the experimental results of Awin and Dukhan [10], it was depicted that by comparing with serpentine flow field, metal foam design can offer more homogeneous distributions of temperature on the MEA. Nevertheless, for liquid-cooled PEMFC, the utilization of metal foam flow field causes water discharge issue affecting the operation stability. Tseng et al. [11] identified that some pores in metal foam were clogged by liquid water, causing the water flooding phenomenon. In contrast, in the case of air-cooled PEMFC, the adoption of a metal foam flow field demonstrates the capability to retain water leading to a better equilibrium between heat dispersion and water preservation. Kang et al. [12] carried out numerical and experimental researches to examine the influence of metal foam on the water management as well as output performance of air-cooled PEMFC. The findings obtained from their research revealed that the implementation of metal foam in the cathode side effectively inhibits the membrane dehydration phenomenon. In view of this, the design strategy for metal foam flow field in various types of PEMFC is crucial for improving the cell performance and enhancing the potential for practical implementation of metal foam flow field. However, the study related to the metal foam design strategy for PEMFC is rare in the open literature.

In this chapter, the design strategy for metal foam flow field in liquid-cooled PEMFC and air-cooled PEMFC is extensively investigated through three sections. The influence of the arrangement of metal foam flow field on the cell performance as well as water management is thoroughly explored. A 3-hour constant current density operational test is conducted for assessing the operation stability associated with the water flooding phenomenon. Structural characteristics of metal foam flow field are also developed for addressing the water flooding issue. Furthermore, the influence of metal foam flow field on the thermal management and performance of air-cooled PEMFC is studied.

2. Experimental

2.1 Design of metal foam flow field and assembled of single cell

The metal foam with various structures applied in this chapter was nickel-based (Guangjiayuan Electronic Materials Co., Ltd., Kunshan City). Since the active area of PEMFC is 50 mm × 50 mm, it is necessary to cut the sample by a laser with a precision of ±0.1 mm to guarantee that the pores located in the cutting region remain unclogged. Additionally, the experimental MEA adopted for this chapter was

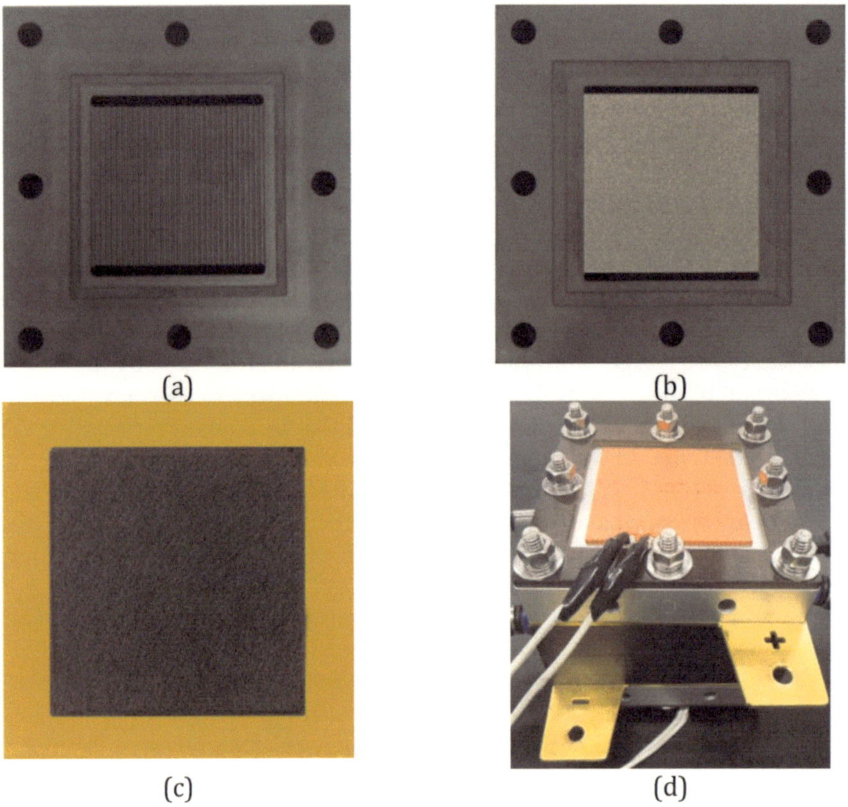

Figure 2.
Straight flow field, metal foam flow field, MEA, and single cell.

produced by Wuhan University of Technology. In **Figure 2**, the BP with different flow fields, MEA, and single cell used for experiments are depicted.

2.2 Test system

The experimental setup for testing PEMFC, referred to as the G60 fuel cell test system, is depicted in **Figure 3**. This system comprises various components like a gas storage device, heating device, pipeline network, moisture control device, thermal insulation device, monitoring and control device, PEMFC, reactants mass flow control device, and so on.

3. Performance and water management investigation on the arrangement of metal foam flow field in liquid-cooled PEMFC

In this section, four cases are established as depicted in **Table 1** to evaluate the influence of the arrangement of metal foam flow field on the performance of PEMFC. The difference in the water management of different cases is also examined. Note that the porosity, thickness, and pore diameter of metal foam employed in this section are 95–97%, 1, and 0.2 mm, respectively. The experimental operation conditions are also depicted in **Table 2**.

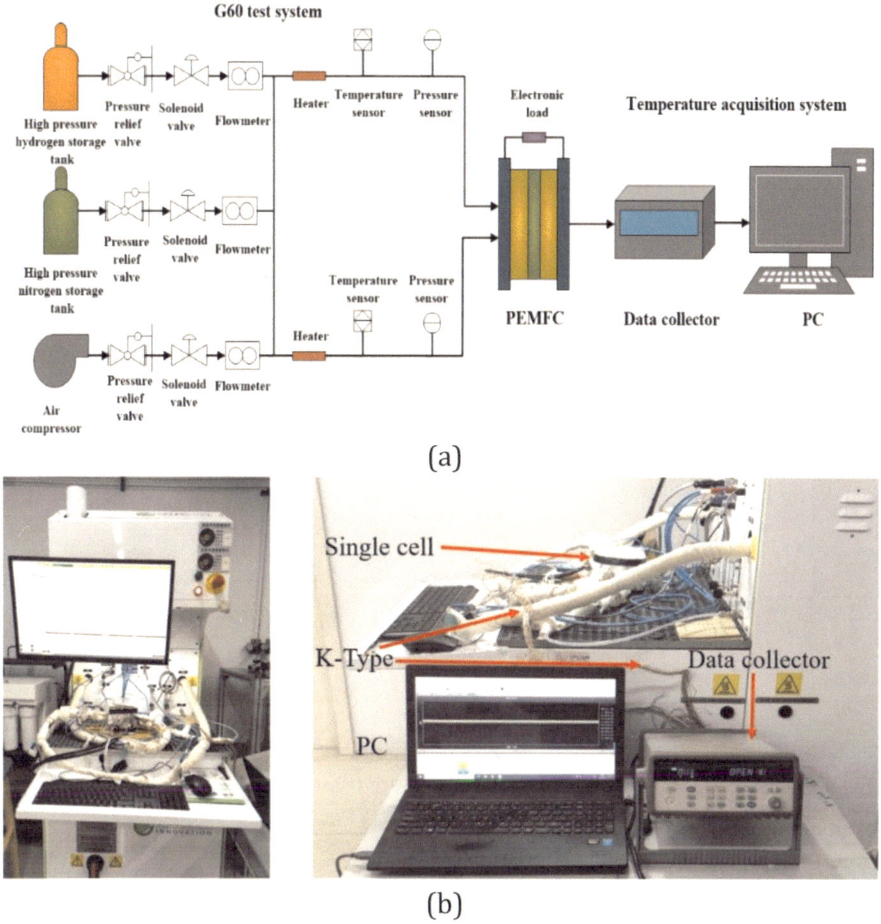

Figure 3.
Diagrammatic representation and picture of the entire experimental system.

Case No.	Explanations
Case 1	Straight flow field for anode and cathode
Case 2	Metal foam flow field for anode and straight flow field for cathode
Case 3	Straight flow field for anode and metal foam flow field for cathode
Case 4	Metal foam flow field for anode and cathode

Table 1.
Explanation of different cases.

3.1 Influence of relative humidity

The influence of relative humidity (RH) of reactants on the cell performance was studied in this subsection; **Figure 4** shows the peak power density of four cases and the peak power density augmentation rate (η) for cases 2–4 in comparison with that for case 1 under various RHs.

Operation condition	Value
Operation temperature	70°C
Relative humidity of H_2	0/0.5/0.8/1
Relative humidity of air	0/0.5/0.8/1
Stoichiometric flow ratio of H_2	1.2
Stoichiometric flow ratio of air	2.5
Operating pressure	1 atm

Table 2.
Operation conditions for experiments in this section.

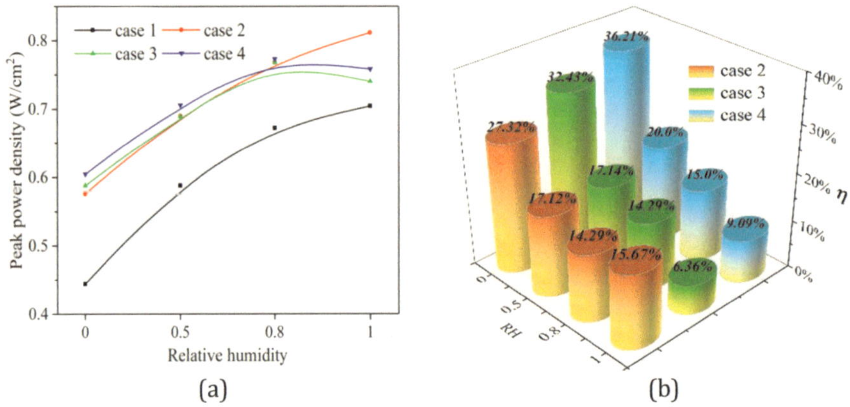

Figure 4.
(a) Peak power density of four cases under various RHs; (b) peak power density augmentation rate of cases 2–4 in comparison with that of case 1.

According to **Figure 4**, adopting metal foam flow field can result in the improved cell performance over traditional straight flow field, which conforms to the findings from Jo and Ju [13]. Additionally, when RH is set to 1, the augmentation rates of peak power density in cases 3 and 4, both utilizing metal foam flow field in the cathode side, are 6.36 and 9.09%, respectively. In case 2, the augmentation rate η is more noticeable, reaching 15.67% when only utilizing metal foam flow field in the anode side. The reason is that liquid water is produced in the cathode catalytic layer during the operation of PEMFC. Furthermore, the migration of protons through the membrane also results in the accumulation of liquid water in the cathode side. Therefore, when setting a high RH, a significant quantity of liquid water can fill the reactants passage in the gas diffusion layer (GDL) in the cathode side, resulting in water flooding and detrimental effects on the performance of PEMFC.

3.2 Flooding test and analysis

According to the study conducted by Bao et al. [14], it was asserted that removing liquid water poses a greater challenge for metal foam flow field in comparison with traditional straight flow field, primarily owing to the complex internal structure impeding effective water removal. When RH and current are high, the implementation of metal foam in the cathode side results in the water flooding phenomenon, affecting the cell performance. In contrast, the implementation of metal foam flow field demonstrates water retention capability when RH is low, wetting the membrane, thereby enhancing the rate of mass transfer in the membrane and improving the cell performance in comparison with traditional straight flow field.

Considering the test processes mentioned above, PEMFC adopting metal foam as cathode flow field easily results in the occurrence of the water flooding phenomenon at high current and RH. In order to gain deeper insights into the flooding phenomenon, 3-hour constant current density operational test on four cases under different RHs was performed in this subsection. The current density is fixed at 1.5 A/cm^2. **Figure 5** presents the variations of voltage and current density over time in different cases under RH = 1. It is evident that over the 3 hours of operation, the voltages and current densities of cases 1 and 2 essentially remain steady. However, there were four minor voltage attenuations during the initial 105 minutes in cases 3 and 4. Following that, the voltage started to attenuate quickly. The underlying cause for this can be attributed to the implementation of metal foam as flow field in the cathode side in these two cases. In the course of PEMFC operation, liquid water slowly accrues and cannot be effectively removed in time, resulting in the reactants channel in the GDL to be filled. The voltage gradually decreased, while PEMFC operates in a constant current density model.

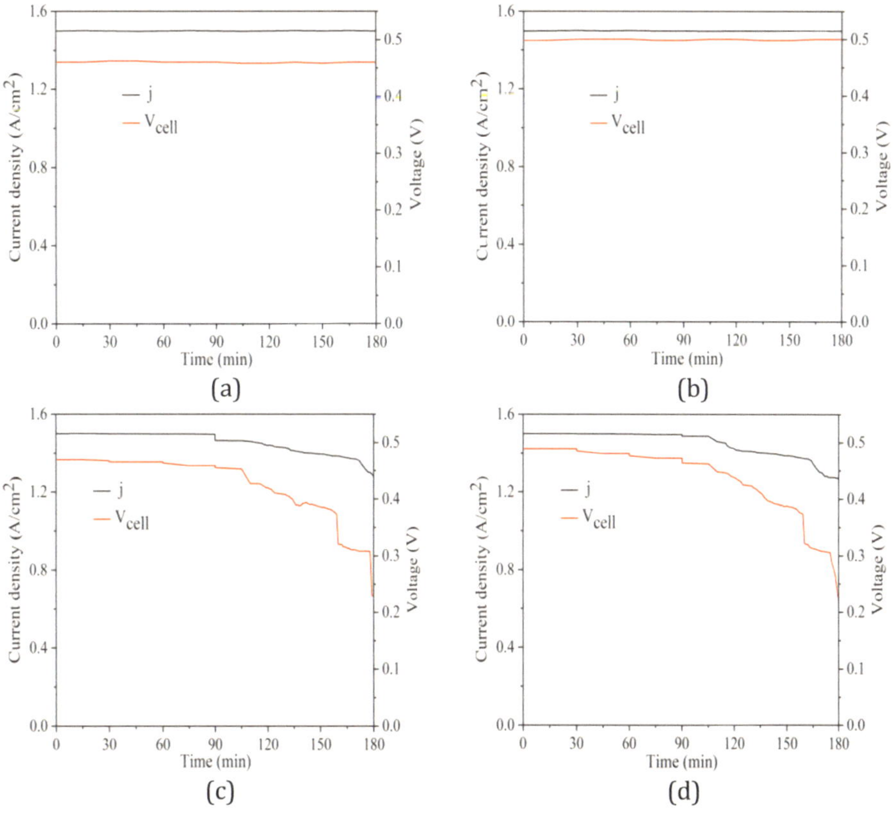

Figure 5.
Voltage and current density over time of different cases under RH of 1. (a) Case 1. (b) Case 2. (c) Case 3. (d) Case 4.

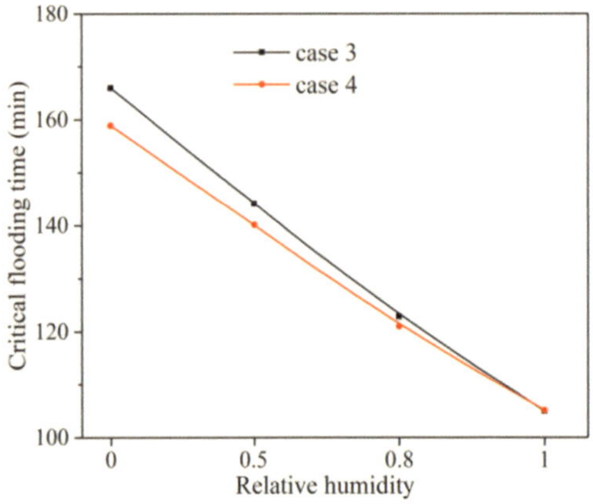

Figure 6.
Correlations between the RH and crucial time of flooding in cases 3 and 4.

In **Figure 6**, the correlations between RH and crucial flooding time of cases 3 and 4 are depicted. It is evident that as RH decreases, the crucial flooding times for these two cases experience a gradual delay. Thus, it can draw the conclusion that arranging metal foam flow field in the cathode side leads to the water flooding issue, negatively impacting the cell performance.

4. Investigation of water management and structure optimization of metal foam flow field in liquid-cooled PEMFC

As stated before, the water flooding phenomenon occurs under the high current density and high RH, when metal foam was adopted in the cathode side of PEMFC. In addition, Bao et al. [15] pointed out that the reactant velocity affected by structural parameters of flow field has a significant influence on the water discharging in PEMFC. Therefore, the influences of the structural characteristics of metal foam like compression rate and pore per inch (PPI) on the water management as well as cell performance are investigated in this section. It should be noted that the cathode flow field depth is 0.5 mm. Traditional straight flow field is set with channel dimensions of 1 mm for width, height, and rib width in the anode side. RHs for both anode and cathode are all 1.

4.1 Influence of compression rate

For the purpose of examining the influence of the metal foam compression rate on the water management and cell performance, metal foam flow fields with various compression rates are set up in the cathode side of PEMFC in this subsection. **Figure 7** displays SEM images of metal foam including various compression rates. Obviously, the pore diameter of metal foam in the direction perpendicular to the reactants flow reduces, while it basically remains unchanged in the flow direction of reactants as the increase of the compression rate. Hence, the reactant diffusion to the GDL can be enhanced through the compression of metal foam. However, compressing metal foam will also lead to a reduction in porosity. Based on the results of Shin et al. [16] and Tseng et al. [11], metal foam has a larger porosity, which indicates that there is more

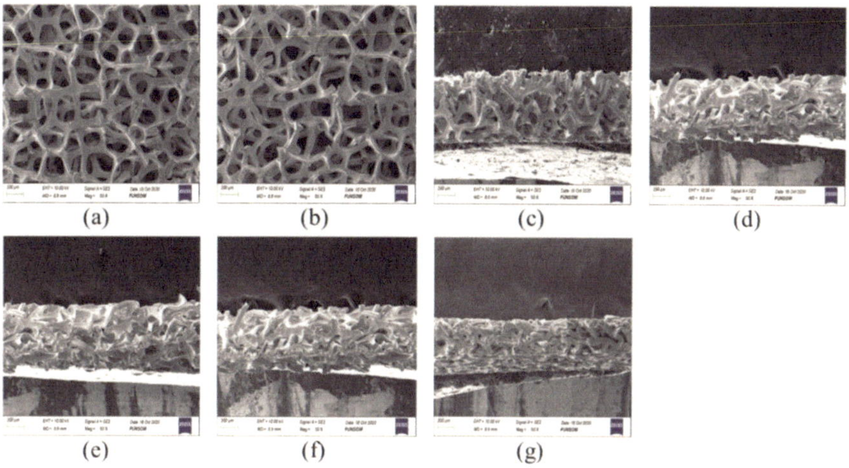

Figure 7.
SEM images: (a, b) overhead perspective of metal foam exhibiting compression rates of 0 and 0.75; (c, g) cross-sectional depiction of metal foams showcasing diverse compression rates, specifically 0, 0.38, 0.50, 0.69, and 0.75.

space for reactants improving the homogeneity of the reactant distribution. Thus, compressing metal foam does not favor achieving homogeneous reactant distribution.

From **Figure 8(a)**, when the compression rate of metal foam flow field in the cathode side rises, it is clear that the cell performance is enhanced. As shown in

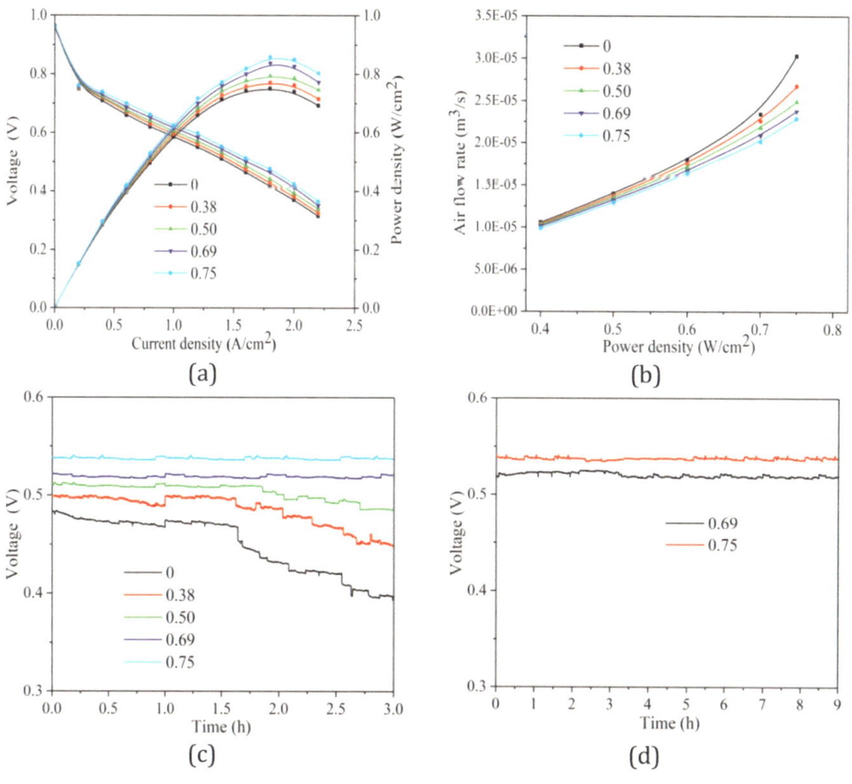

Figure 8.
(a) Polarization curves; (b) comparison of air flow rate; (c, d) voltage over time.

Figure 8(b), the cell performance enhancement also improves the air utilization. The reason is that reactant diffusion to the CL is boosted by the increment in the compression rate of metal foam flow field. This notion is also supported by the findings of Park et al. [17]. More crucially, as demonstrated in **Figure 8(c)**, the increment in the compression rate also leads to better voltage stability of PEMFC, suggesting that the water management capability is enhanced. It is worth noting that the operation voltage is totally stable as soon as the metal foam compression rate rises over 0.69. In order to analyze this point more clearly, the duration of constant current density operational tests for PEMFCs utilizing the compression rates of 0.69 and 0.75 was extended to 9 hours as depicted in **Figure 8(d)**. Based on the results presented in **Figure 8(c, d)**, the utilization of metal foam with a compression rate higher than 0.69 can significantly improve the water management in PEMFCs.

4.2 Influence of PPI

In this subsection, the cathode of PEMFC is equipped with metal foam flow fields featuring various PPIs. From **Figure 9**, the pore density rises as the increment of PPI of metal foam, which is helpful for the reactants to be distributed uniformly. However, the increase of PPI will also lead to a reduction in pore diameter and permeability of metal foam. Liquid water first percolates into metal foam flow field *via* the GDL and subsequently exits through flow field outlet. The removal of liquid water through the GDL to cathode flow field is hindered by the reduction in permeability. As a result, the synergistic influence of PPI on homogeneity of reactants distribution and water management is examined in this subsection.

The polarization curves of PEMFCs comprising various PPIs in the cathode side are depicted in **Figure 10(a)**. It is interesting to observe that the performance of PEMFC adopting metal foam flow field with 75 PPI is the best. This outcome can be attributed to the synergistic influence between reactant distribution and water management. In **Figure 10(b, c)**, the pressure drop from cathode inlet to outlet and operation voltage stability of PEMFCs are provided. As seen in **Figure 10(c)**, under current density of 2.2 A/cm^2, the pressure drops from cathode inlet to outlet of PEMFCs with 35 PPI, 75 PPI, and 110 PPI are 17.2, 25.1, and 27.8 kPa, respectively. It is also worth noting that prior to 1.25 A/cm^2, the pressure drop of metal foam with 110 PPI is practically indistinguishable from that with 75 PPI. Revealing that the disparity in the pressure drops attributable to pore diameter between 75 PPI and 110 PPI can be disregarded in the presence of a small amount liquid water. However, the pressure drop of 110 PPI is

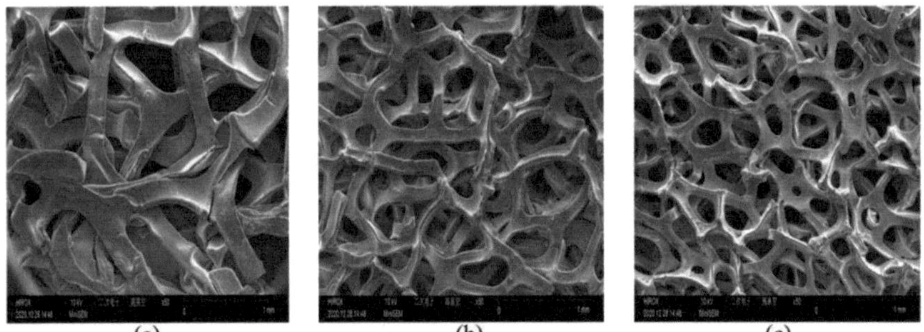

Figure 9.
SEM images: Overhead perspective of metal foams with various PPIs: (a) 35 PPI, (b) 75 PPI, (c) 110 PPI.

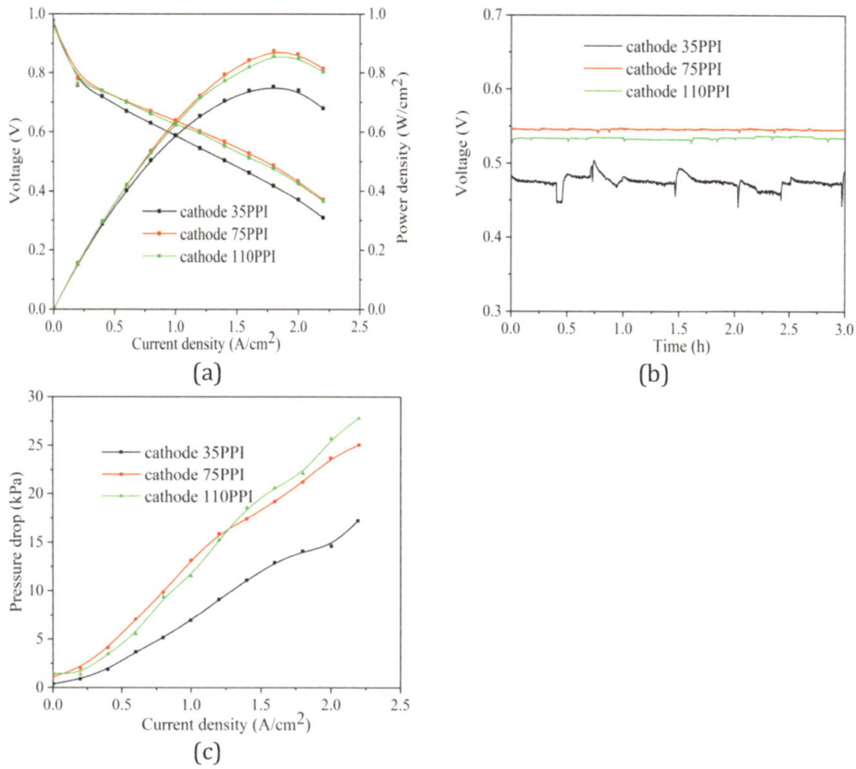

Figure 10.
(a) Polarization curves; (b) voltage over time; (c) pressure drops from cathode inlet to outlet under different current densities.

obviously larger than that of 75 PPI, after 1.25 A/cm^2, owing to the fact that more liquid water was generated as the increase of current density. The low permeability of 110 PPI metal foam leads to lower drainage efficiency for liquid water compared to 75 PPI metal foam. The 3-hour operational stability tests under current density of 1.5 A/cm^2 as depicted in **Figure 10(b)** also indicate this point. For 35 PPI metal foam flow field, the obvious voltage fluctuation was observed. The voltage stability was improved with the further rise of PPI, proving the improved water management capability.

5. Thermal management enhancement of air-cooled PEMFC through the implementation of metal foam flow field

In this section, the influence of metal foam flow field on the thermal management and performance of air-cooled PEMFC is examined, considering the superior heat dissipation and water retention characteristics of metal foam. Straight flow field is implemented for the anode side, the channel sizes of which are identical to those mentioned in the preceding sections. **Table 3** provides a comprehensive list of the geometric parameters related to the cathode flow field. In addition, the experimental operation conditions are also provided in **Table 4**.

Case No.	Geometric parameters of cathode flow field		Flow field type
	Length/width/height of channel (mm)	Width of rib (mm)	
Case 5	50/1/1	1	Straight flow field
Case 6	50/1/1	1	Metal foam flow field
Case 7	50/3/1	1	Straight flow field
Case 8	50/3/1	1	Metal foam flow field
Case 9	50/5/1	1	Straight flow field
Case 10	50/5/1	1	Metal foam flow field
Case 11	50/5/2	1	Metal foam flow field
Case 12	50/5/3	1	Metal foam flow field

Table 3.
Descriptions of experiment cases in this section.

Operation condition	Value
Inlet temperature of H_2	24°C
Inlet temperature of air	24°C
Relative humidity of inlet H_2	0%
Relative humidity of inlet air	50%
Stoichiometric flow ratio of H_2	1.5
Stoichiometric flow ratio of air	50
Operating pressure (gauge pressure)	1 atm

Table 4.
Operation conditions for the experiments in this section.

5.1 Thermal performance evaluation

As air-cooled PEMFC operates, electricity (Q_{ele}) and heat (Q_{he}) are the products of the electrochemical reaction process. The main ways for heat dissipation are listed as following: (1) forced convective heat transfer of air (Q_1), it is worth observing that forced convective heat transfer of hydrogen can be ignored because of its low flow rate in comparison to air; (2) the cell temperature rise (Q_2); (3) thermal loss resulting from natural convective heat transfer as well as thermal radiation of PEMFC surface (Q_3). In there, Q_3 can be ignored through using the heat-insulating material to pack the air-cooled PEMFC; thermal conductivity of this insulating material is 0.03 W/(m^2/K). As a result, forced convective heat transfer of air and the rise of cell temperature are the main topics in this section. Q_{ele}, Q_{he}, and Q_1 are described as following definitions [18]:

$$Q_{ele} = I \cdot V \tag{1}$$

$$Q_{he} = I(E_r n - V) = Q_1 + Q_2 + Q_3 \tag{2}$$

$$Q_1 = q_{air} \rho c_p (T_{out} - T_{in}) \tag{3}$$

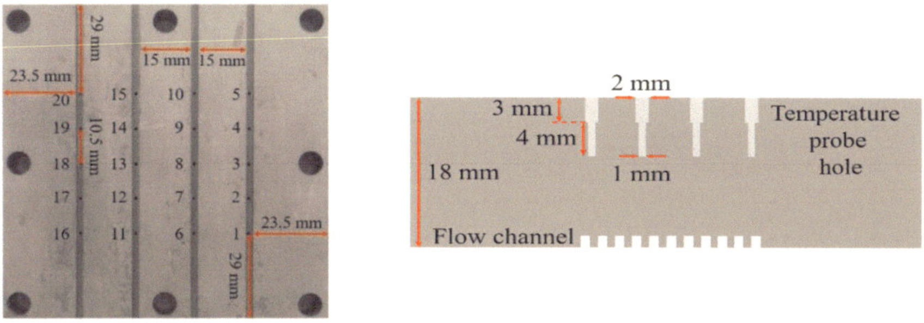

Figure 11.
Exact locations of thermocouples.

where E_r is designated as the theoretical reversible voltage of PEMFC (usually around 1.254 V), n is referred as the quantity of cells, q_{air} is referred as the air flow rate (m^3/s), ρ is designated as air density (kg/m^3), and c_p is referred as the air specific heat (J/(kg·K)).

The temperature acquisition system was adopted to gather the temperature information of PEMFC. A total of 20 temperature detection points were arranged as depicted in **Figure 11**. The mean cell temperature is defined as the average temperature of 20 thermocouples. The maximum temperature difference (ΔT_{max}) and temperature uniformity index (TUI) are adopted to characterize the temperature distribution uniformity. The cooling efficiency (ε_c) of PEMFC can be determined through utilizing the proportion of Q_1 to Q_{he} [19]. ΔT_{max}, TUI, and ε_c are expressed as below:

$$\Delta T_{\max} = T_{\max} - T_{\min} - \Delta T_{ini} \tag{4}$$

$$T_{ave} = \frac{\Sigma_{n=1}^{N} T_n}{N} \tag{5}$$

$$TUI = \frac{\Sigma_{n=1}^{N} |T_n - T_{ave}|}{N} \tag{6}$$

$$\varepsilon_c = \frac{Q_1}{Q_{he}} \tag{7}$$

Note that T_{max} and T_{min} are designated as the maximum and minimum temperatures recorded by 20 thermocouples, T_{ave} is referred as the average temperature calculated from these 20 thermocouples, and ΔT_{ini} is used to define the initial temperature difference of PEMFC before the start of the experiment that consistently remains below 0.2°C.

5.2 Influence of metal foam flow field on thermal management of air-cooled PEMFC

In this subsection, two cases, namely, cases 5 and 6, are chosen to examine the influence of metal foam flow field on the thermal management of air-cooled PEMFC. According to the data from **Figure 12(a, b)**, case 6 gains a greater cooling efficiency in comparison with case 5. The primary cause comprises the following two factors: (1) The significant mixing effect of metal foam leads to an improvement in the forced

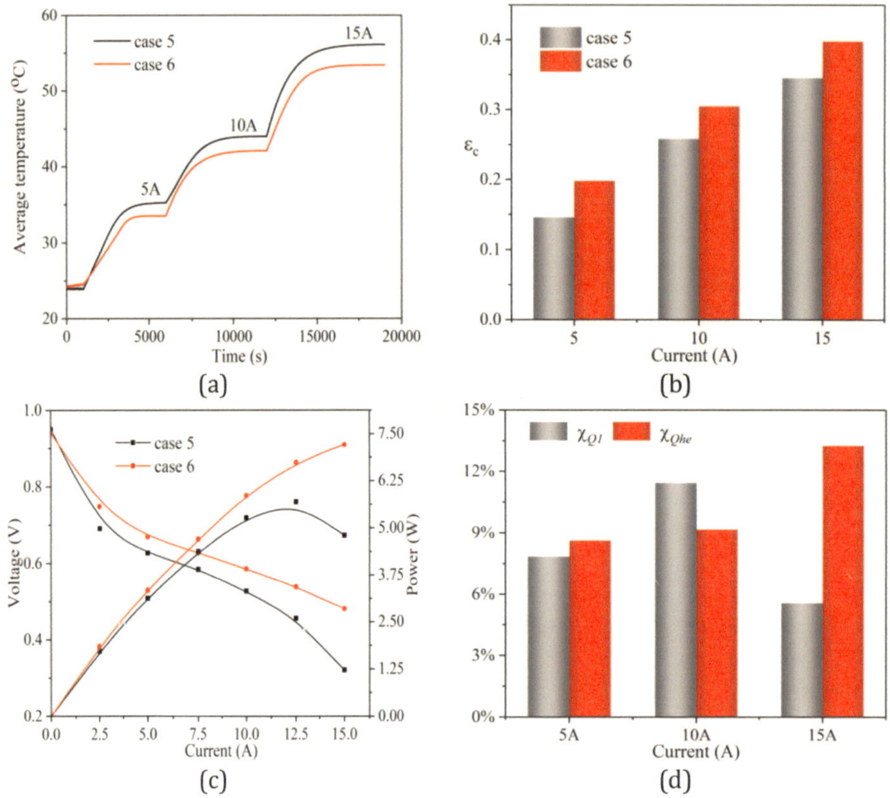

Figure 12.
(a) Cell temperature over time; (b) comparison of the cooling efficiency; (c) polarization curves; (d) variation rates of Q_1 *and* Q_{he} *of case 6 in comparison with those of case 5.*

convective heat transfer in fuel cell; (2) according to **Figure 12(c)**, the output voltage of case 6 is greater in comparison with that of case 5 particularly when the operation current is located in medium and high regions; this observation suggests a notable reduction in ohmic loss. The explanation for this phenomenon lies in a better capability to retain water of metal foam, which has a positive impact on the proton transport efficiency of the proton exchange membrane. The results of Peng et al. [20] also emphasized this correlation. The improvement in the performance of PEMFC indicates a decrease in the heat produced from the electrochemical reaction. For conducting a more thorough examination of the influences of these two elements on the thermal management of air-cooled PEMFC, in comparison with case 5, the enhancement rate in forced convective heat transfer (χ_{Q1}) and the decrease rate in the generated heat (χ_{Qhe}) for case 6 are depicted in **Figure 12(d)**. The computing means for χ_{Q1} and χ_{Qhe} are presented in Eq. (8). It is worth noting that in the case of current below 10 A, there is minor variance between χ_{Q1} and χ_{Qhe}. However, χ_{Qhe} is significantly larger than χ_{Q1} under current of 15 A, meaning that at high current, employing metal foam flow field effectively reduces the heat generated from the electrochemical reaction process. The findings are in line with the outcomes introduced by Kang et al. [12]. Based on **Figure 13(a, b)**, it can be observed that in comparison with case 5, case 6 has smaller ΔT_{max} and TUI, demonstrating that applying metal foam in the cathode side can help homogenize the temperature distribution of PEMFC, owing to the significant mixing effect.

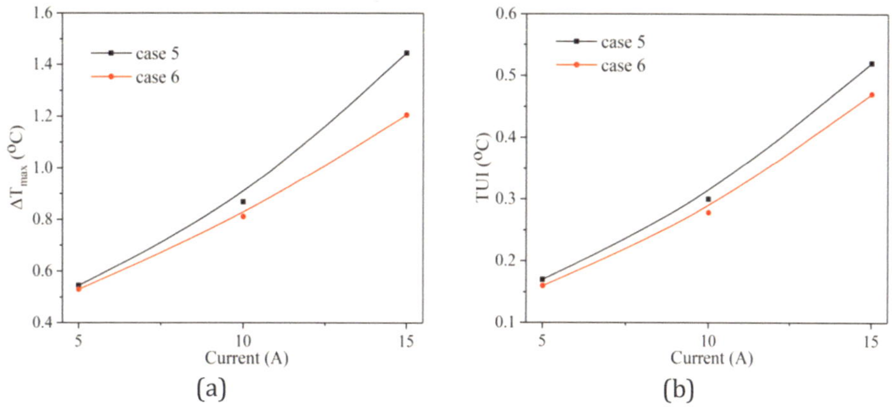

Figure 13.
(a, b) temperature uniformities of different cases.

$$\chi_{Q1/Qhe} = \frac{\left|\beta_6^{Q1/Qhe} - \beta_5^{Q1/Qhe}\right|}{\beta_5^{Q1/Qhe}} \times 100\% \quad (8)$$

where $\beta^{Q1/Qhe}$ is defined as the value of forced convective heat transfer or the heat produced from the electrochemical reaction process.

5.3 Influence of width of metal foam flow field

In this subsection, the study focused on investigating the influence of the width of metal foam on the thermal management of air-cooled PEMFC. Cases 6, 8, and 10 are analyzed. **Figure 14(a, b)** depict the average temperatures and cooling efficiencies of three cases at various currents. The results clearly demonstrate that case 10 outperforms the other two cases in terms of the thermal management. When raising the width from 1 mm (case 6) to 5 mm (case 10) with a current of 15 A, the average temperature of air-cooled PEMFC decreases by 8.4°C and the cooling efficiency increases by 14.7%, which means that the thermal management capability can be boosted through raising the width of metal foam. Furthermore, when maintaining a fixed rib width of 1 mm, raising the width of metal foam enlarges the contact area between the MEA and metal foam flow field, resulting in an improvement in the distribution uniformity of air as well as cell performance. As a result, less heat is generated from electrochemical reaction, as supported by the polarization curves depicted in **Figure 14(c)**. In addition, the rise of the metal foam width also results in more even distribution of air on the MEA, leading to the homogeneity enhancement of the temperature distribution as depicted in **Figure 15(a, b)**.

5.4 Influence of height of metal foam flow field

In this subsection, cases 10, 11, and 12 are chosen for analyzing the influence of the height of metal foam on the thermal management of air-cooled PEMFC. Observing from **Figure 16(a, b)**, it is revealed that when current is lower than 10 A, case 12 exhibits the lowest cooling efficiency compared to cases 10 and 11. However, when current is further increased from 10 A to 15 A, case 12 experiences a significant increase in the cooling efficiency and becomes comparable to cases 10 and 11. In order

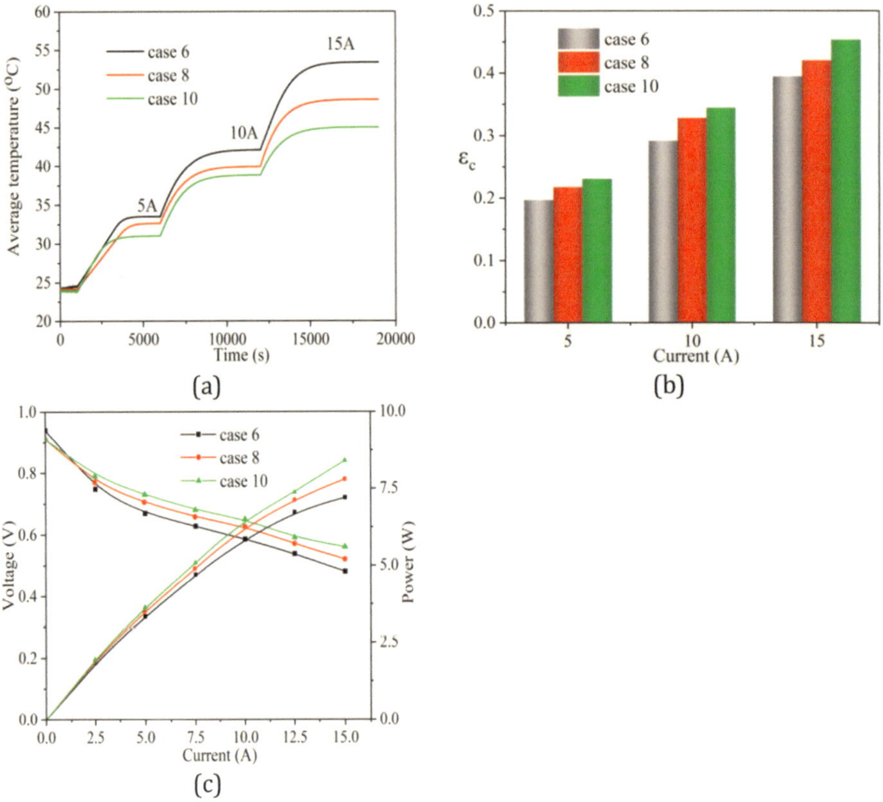

Figure 14.
(a) Cell temperature over time; (b) comparison of the cooling efficiency; (c) polarization curves.

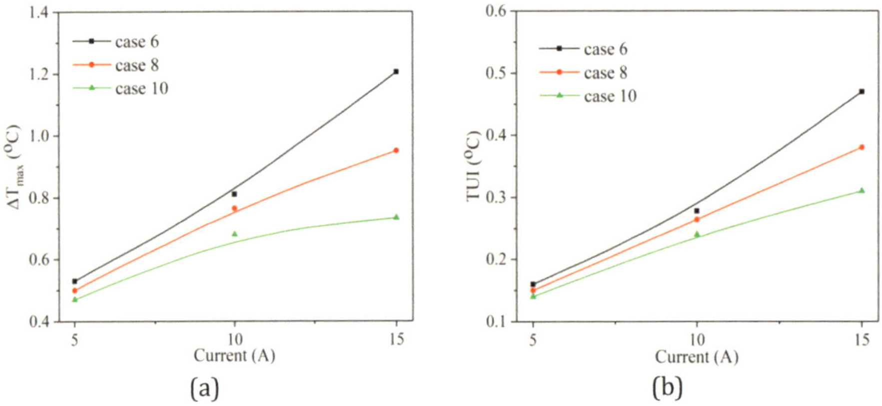

Figure 15.
(a, b) temperature uniformities of different cases.

to exemplify this point, **Figure 16(c, e)** are provided. From **Figure 16(c)**, case 12 has the lowest performance, resulting in more waste heat generation. The operation voltage stabilities of different cases are depicted in **Figure 16(d)**. It is observed that when current is lower than 10 A, case 12 exhibits instability. Nevertheless, the operation stability of case 12 experiences a significant enhancement with a current of 15 A.

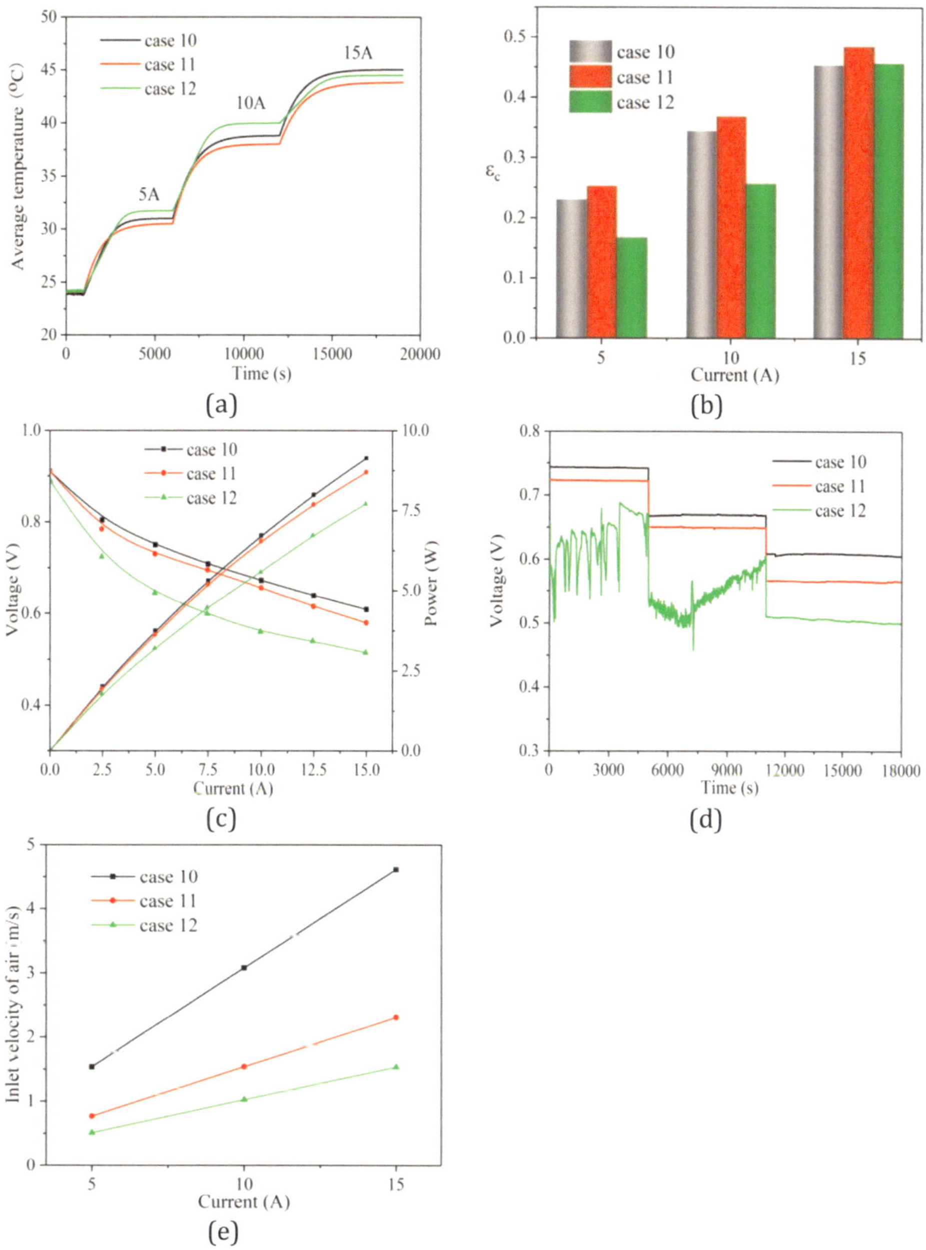

Figure 16.
(a) Cell temperature over time; (b) comparison of the cooling efficiency; (c) polarization curves; (d) voltage over time; (e) air inlet velocities of three cases.

The primary factor lies in the fact that in comparison to case 10 and case 11, the air inlet velocity of case 12 is lower as illustrated in **Figure 16(e)**. In the cases where current is lower than 10 A, the generated water from electrochemical reaction in case 12 cannot be effectively discharged by the airflow, leading to the operation instability. However, at current of 15 A, the increased inlet velocity of air allows for efficient water discharge, leading to the improved operation stability as well as output voltage in case 12. Furthermore, case 12 exhibits the best mixing effect in comparison with case 10 and case 11. As a result, the cooling efficiency of case 12 experiences a significant enhancement at current of 15 A. Among the three cases, case 11 shows the highest cooling efficiency, suggesting that for case 11, the benefits of raising the height of metal foam outweigh potential unfavorable consequences.

6. Conclusion

This chapter focuses on the experimental investigation of the design strategy for metal foam flow field in different types of PEMFC. The influence of the arrangement of metal foam flow field on the performance and water management of liquid-cooled PEMFC was experimentally examined. Additionally, the structural characteristics of metal foam were optimized to enhance the water management and cell performance. Furthermore, in air-cooled PEMFC, the influence of metal foam flow field on the thermal management was also explored. On the basis of the exhausted experiments, the following conclusions have been drawn:

1. Under low RH, metal foam flow field demonstrates the capability to retain water, which is essential for wetting the membrane and enhancing the cell performance. Conversely, under high RH, employing metal foam flow field in the cathode side increases the possibility of water flooding, thereby impacting the liquid-cooled PEMFC performance negatively. Interestingly, when only the anode side utilizes metal foam flow field, there is no instance of the water flooding phenomenon, leading to an improved cell performance.

2. Under the operating conditions with complete relative humidity and a constant current density of 1.5 A/cm^2, liquid-cooled PEMFC utilizing traditional straight flow field in the cathode side demonstrates steady current density and voltage throughout a 3-hour operation. However, when utilizing metal foam flow field in the cathode side, PEMFC experienced slight flooding within 105 minutes, leading to a slight degradation in voltage. Subsequently, the voltage decayed quickly. Furthermore, as relative humidity decreases, the crucial flooding times are gradually delayed.

3. Performance enhancement in liquid-cooled PEMFCs is observed with an increase in metal foam compression rate in the cathode side. Additionally, once the compression rate in the cathode side exceeds 0.69, the operation voltage demonstrates complete stability during 3-hour operation.

4. Regarding PPI of metal foam, it is intriguing to observe that optimal PPI, namely, 75, is obtained for the cathode side. This is attributed to the synergistic influence between the water management and gas distribution of metal foam flow fields with various PPIs.

5. The implementation of metal foam as cathode flow field for air-cooled PEMFC can effectively enhance its thermal management capability. Moreover, the implementation of metal foam can help even out the temperature distribution owing to its mixing effect for reactants.

6. Enhancing the width of metal foam results in better thermal management for air-cooled PEMFC. In the case where the height is fixed at 1 mm and the width rises

from 1 to 5 mm, the cell temperature lowers by 8.4°C, accompanied by a 14.7% increase in the cooling efficiency at current of 15 A.

7. When the width of metal foam flow field is fixed at 5 mm, raising the height of metal foam from 1 to 2 mm results in a slight improvement in the cooling efficiency of air-cooled PEMFC. However, a further increase in the height to 3 mm leads to a minor water flooding phenomenon due to the low inlet velocity under current below 10 A.

Acknowledgements

This work is supported by National Natural Science Foundation of China (Nos. 52276184, 52176062, 51976055, 52076072) and the Science and Technology Innovation Program of Hunan Province (2020RC4040, 2021GK2017).

Nomenclature

Q_{ele}	electric energy (W)
Q_{he}	heat (W)
Q_1	forced convective heat transfer (W)
Q_2	cell temperature rise (°C)
I	current (A)
V	voltage (V)
E_r	the theoretical reversible voltage (V)
n_{cell}	number of cells (−)
q_{air}	flow rate of air (m^3/s)
T	temperatures (K)
ΔT_{ini}	initial temperature difference (°C)
TUI	temperature uniformity index (°C)
c_p	specific heat of air (J/(kg·K))

Greek letters

η	maximum power density enhancement
ρ	density of air (kg/m^3)

Subscripts and superscripts

max	maximum
min	minimum
ave	average
ini	initial
he	heat
in	inlet
out	outlet

Author details

Zhongmin Wan[1], Yun Sun[2] and Chen Yang[2*]

1 College of Mechanical Engineering, Hunan Institute of Science and Technology, Yueyang, China

2 Fujian Universities Engineering Research Center of Reactive Distillation Technology, College of Chemical Engineering, Fuzhou University, Fuzhou, Fujian, China

*Address all correspondence to: cyang@fzu.edu.cn

References

[1] Barelli L, Bidini G, Ottaviano A. Optimization of a PEMFC/battery pack power system for a bus application. Applied Energy. 2012;**97**:777-784. DOI: 10.1016/j.apenergy.2011.11.043

[2] Kang S, Min K. Dynamic simulation of a fuel cell hybrid vehicle during the federal test procedure-75 driving cycle. Applied Energy. 2016;**161**:181-196. DOI: 10.1016/j.apenergy.2015.09.093

[3] Wan Z, Quan W, Yang C, Yan H, Chen X, Huang TM, et al. Optimal design of a novel M-like channel in bipolar plates of proton exchange membrane fuel cell based on minimum entropy generation. Energy Conversion and Management. 2020;**205**:112386. DOI: 10.1016/j.enconman.2019.112386

[4] Cai G, Liang Y, Liu Z, Liu W. Design and optimization of bio-inspired wave like channel for a PEM fuel cell applying genetic algorithm. Energy. 2020;**192**: 116670. DOI: 10.1016/j.energy.2019.116670

[5] Chen H, Guo H, Ye F, Ma CF. Modification of the two-fluid model and experimental study of proton exchange membrane fuel cells with baffled flow channels. Energy Conversion and Management. 2019;**195**:972-988. DOI: 10.1016/j.enconman.2019.05.071

[6] Yan X, Guan C, Zhang Y, Jiang K, Wei G, Cheng X, et al. Flow field design with 3D geometry for proton exchange membrane fuel cells. Applied Thermal Engineering. 2019;**147**:1107-1114. DOI: 10.1016/j.applthermaleng.2018.09.110

[7] Yang C, Nakayama A. A synthesis of tortuosity and dispersion in effective thermal conductivity of porous media. International Journal of Heat and Mass Transfer. 2010;**53**:3222-3230. DOI: 10.1016/j.ijheatmasstransfer.2010.03.004

[8] Nakayama A, Ando K, Yang C, Sano Y, Kuwahara F, Liu J. A study on interstitial heat transfer in consolidated and unconsolidated porous media. Heat and Mass Transfer. 2009;**45**(11): 1365-1372. DOI: 10.1007/s00231-009-0513-x

[9] Afshari E, Mosharaf-Dehkordi M, Rajabian H. An investigation of the PEM fuel cells performance with partially restricted cathode floe channels and metal foam as a flow distributor. Energy. 2017;**118**:705-715. DOI: 10.1016/j.energy.2016.10.101

[10] Awin Y, Dukhan N. Experimental performance assessment of metal-foam flow fields for proton exchange membrane fuel cells. Applied Energy. 2019;**252**:113458. DOI: 10.1016/j.apenergy.2019.113458

[11] Tseng CJ, Tsai BT, Liu ZS, Cheng TC, Chang WC, Lo SK. A PEM fuel cell with metal foam as flow distributor. Energy Conversion and Management. 2012;**62**:14-21. DOI: 10.1016/j.enconman.2012.03.018

[12] Kang DG, Park C, Lim IS, Choi SH, Lee DK, Kim MS. Performance enhancement of air-cooled open cathode polymer electrolyte membrane fuel cell with inserting metal foam in the cathode side. International Journal of Hydrogen Energy. 2020;**45**:27622-27631. DOI: 10.1016/j.ijhydene.2020.07.102

[13] Jo A, Ju H. Numerical study on applicability of metal foam as flow distributor in polymer electrolyte fuel cells (PEFCs). International Journal of Hydrogen Energy. 2018;**43**(30):

14012-14026. DOI: 10.1016/j.ijhydene.2018.01.003

[14] Bao Z, Niu Z, Jiao K. Numerical simulation for metal foam two-phase flow field of proton exchange membrane fuel cell. International Journal of Hydrogen Energy. 2019;**44**(12): 6229-6244. DOI: 10.1016/j.ijhydene.2019.01.086

[15] Bao Z, Wang Y, Jiao K. Liquid droplet detachment and dispersion in metal foam flow field of polymer electrolyte membrane fuel cell. Journal of Power Sources. 2020;**480**:229150. DOI: 10.1016/j.jpowsour.2020.229150

[16] Shin DK, Yoo JH, Kang DG, Kim MS. Effect of cell size in metal foam inserted to the air channel of polymer electrolyte membrane fuel cell for high performance. Renewable Energy. 2018;**115**:663-675. DOI: 10.1016/j.renene.2017.08.085

[17] Park JE, Hwang WC, Lim MS, Kim SJ, Ahn CY, Kim OH, et al. Achieving breakthrough performance caused by optimized metal foam flow field in fuel cells. International Journal of Hydrogen Energy. 2019;**44**: 22074-22084. DOI: 10.1016/j.ijhydene.2019.06.073

[18] Zhao J, Huang ZP, Jian BX, Bai XY, Jian QF. Thermal performance enhancement of air-cooled proton exchange membrane fuel cells by vapor chambers. Energy Conversion and Management. 2020;**213**:112830. DOI: 10.1016/j.enconman.2020.112830

[19] Mohamed WANW, Talib SFA, Zakaria IA, Mamat AMI, Duad WRW. Effect of dynamic load on the temperature profiles and cooling response time of a proton exchange membrane fuel cell. Journal of the Energy Institute. 2018;**91**(3):349-357. DOI: 10.1016/j.joei.2017.02.006

[20] Peng M, Chen L, Zhang RY, Xu WQ, Tao WQ. Improvement of thermal and water management of air-cooled polymer electrolyte membrane fuel cells by adding porous media into the cathode gas channel. Electrochimica Acta. 2022;**412**:140154. DOI: 10.1016/j.electacta.2022.140154

Chapter 5

Perspective Chapter: Pore Structure Evolution of Cement and Concrete Induced by CO_2 Carbonation

Liwei Zhang, Manguang Gan and Quan Xue

Abstract

This chapter explores the profound impact of CO_2 carbonation on the pore structure of cement and concrete. Carbonation, resulting from the reaction of CO_2 with calcium hydroxide and calcium silicate hydrate in cement, is a complex process that affects the durability and performance of cement-based materials. In this chapter, the changes in pore structure of cement and concrete induced by carbonation are examined. Under low pressure CO_2, cement carbonation leads to the formation of dense carbonation regions. For wellbore cement exposed to high pressure and high concentration of CO_2, a calcite precipitation layer with very low porosity is formed in cement, and two highly porous layers due to dissolution of calcite and cement hydration products are formed at both sides of the calcite precipitation layer. For concrete exposed to atmospheric pressure CO_2, carbonation causes hydration products to form $CaCO_3$ and precipitate in pores. Different from wellbore cement, no distinct calcite precipitation layer is formed in concrete. However, for concrete exposed to 1 MPa CO_2, excessive accumulation of $CaCO_3$ eventually leads to expansion and cracking of pores, which causes the compressive strength of concrete to decrease after reaching the peak.

Keywords: CO_2 carbonation, pore structure, cement, concrete, durability

1. Introduction

The production and application of cement have a history spanning thousands of years. Cement, as a versatile material, has played a pivotal role in the development of civilizations, transforming architectural possibilities and shaping the built environment. The roots of cement can be traced back to ancient times when early civilizations utilized naturally occurring materials to bind structures together. In ancient Egypt, a rudimentary form of cement was created by mixing gypsum with lime. This mixture, known as plaster of paris, was used for construction purposes, including decorative applications and the bonding of stones [1]. Similarly, the Romans utilized a mixture of volcanic ash and lime, known as pozzolana, to construct enduring structures like temples and aqueducts [2].

The next significant advancement in cement technology occurred during the eighteenth and nineteenth centuries with the discovery of hydraulic cement. Before the eighteenth and nineteenth centuries, the main components of cement were lime, volcanic ash, gypsum, and other materials that stuck together when in contact with air. In 1756, British engineer John Smeaton formulated a new type of cement by combining limestone and clay, creating a material that hardened when in contact with water [3]. Smeaton's hydraulic cement was a significant improvement over previous versions, providing increased durability and versatility. In 1796, James Parker produced hydraulic cement by burning clayey limestone. Subsequently, in 1824, Joseph Aspdin obtained a patent for "Portland cement," inspired by the resemblance of his hardened cement to Portland stone. The invention of Portland cement marked the birth of a mature and scalable cementitious material [4]. Portland cement became immensely popular due to its ability to set and harden in wet conditions, making it ideal for various construction applications. The Industrial Revolution brought substantial advancements in cement production. In the mid-nineteenth century, the introduction of rotary kilns enabled the production of larger amounts of cement, fueling the expansion of infrastructure projects worldwide. In 1843, Isaac Charles Johnson, an English engineer, patented the process of adding gypsum to cement, further enhancing its strength and stability.

The turn of the twentieth century witnessed significant breakthroughs in cement production. In 1908, Henry-Louis Le Chatelier, a French chemist, developed the soundness test for cement, ensuring its quality and uniformity. Furthermore, advancements in grinding technology allowed for the production of finely ground cement, increasing its reactivity and setting time. In the early twentieth century, the development of more efficient manufacturing processes revolutionized cement production. These advancements led to the creation of modern Portland cement, which became the primary building material for the construction industry worldwide. Additionally, the advent of blended cements, such as fly ash and slag-based cements, further diversified the range of cement types available [3].

In recent years, the cement industry has witnessed a growing emphasis on sustainability and environmental responsibility. Researchers and manufacturers have been actively exploring alternative materials, such as supplementary cementitious materials (SCMs), to reduce the carbon footprint of cement production [5]. SCMs, including fly ash, slag, and silica fume, can partially replace traditional cement clinker, resulting in lower greenhouse gas emissions. Furthermore, advancements in cement technology have focused on improving the durability and strength of concrete. The development of high-performance concrete, self-healing concrete, and nano-engineered cementitious composites (NECs) has expanded the possibilities of construction, offering increased resistance to extreme conditions and minimizing maintenance requirements [6].

Cement and concrete stand as foundational elements in modern construction and infrastructure, playing an indispensable role in shaping the built environment. The applications of cement and concrete are vast and diverse, ranging from towering skyscrapers and expansive bridges to residential homes and intricate pavements. The durability, versatility, and cost-effectiveness of concrete have made it the backbone of urban development, providing stability and longevity to structures worldwide [7]. Cement and concrete are extensively used in constructing buildings, from residential homes and office complexes to industrial facilities and commercial spaces. They provide the structural integrity required for tall skyscrapers, bridges, dams, and other large-scale structures. Concrete is a preferred material for building highways, roads, and pavements due to its durability and resistance to heavy traffic loads. It offers a smooth and long-lasting surface for transportation networks. Many iconic bridges and tunnels are constructed using reinforced concrete due to its ability to withstand various environmental factors, such as water, corrosion, and seismic activity. Concrete is used in building dams, reservoirs, and other water-related infrastructure. Its impermeability helps in water containment and control, while its strength can withstand pressure from water. Cement has been widely used to construct wellbores for oil and gas exploitation purposes. Most wellbores (especially deep wells) need to be cemented to protect the casing and tubing from corrosion, as well as to prevent fluid migration along the exterior of wellbores [7]. For airport construction, concrete runways, taxiways, and aprons are crucial components of airports. Similarly, port facilities, including wharfs and docks, rely on concrete for their construction. Concrete is used for foundations, walls, floors, and roofs in housing construction. It offers fire resistance, thermal insulation, and noise reduction, contributing to safe and comfortable living spaces. Precast concrete elements, such as beams, columns, and panels, are manufactured off-site and assembled on-site, expediting construction processes. Stamped concrete, stained concrete, and various decorative finishes allow for artistic expression in both interior and exterior architectural designs. Concrete is used to build water treatment plants, reservoirs, and storage tanks, ensuring clean and safe water supply to communities. Nuclear power plants, wind turbine foundations, and electrical substations often rely on concrete for their construction due to its stability and resilience. Stadiums, arenas, and entertainment venues utilize concrete for their grandstands, seating areas, and structural components. Concrete can be employed in coastal protection structures, erosion control systems, and stormwater management infrastructure. Sidewalks, plazas, and public squares often use concrete for pedestrian pathways and outdoor seating areas. Concrete is used for building silos, barns, and other agricultural structures. Concrete has also been widely used in the construction of military bases, bunkers, and defensive structures. These applications showcase the indispensable role that cement and concrete play in shaping the modern world and meeting the diverse needs of society [8].

Portland cement is produced by heating a blend of limestone and clay, or similar materials with comparable compositions, to an ultimate temperature of approximately 1450°C. During this process, the blend fuses, leading to the formation of clinker nodules. The typical composition of clinker is approximately 67% calcium oxide (CaO), 22% silicon dioxide (SiO_2), 5% aluminum oxide (Al_2O_3), 3% ferric oxide (Fe_2O_3), and 3% other components [9]. This clinker generally comprises four main phases: alite, belite, aluminate, and ferrite. Among these, alite (abbreviated as C_3S) holds the utmost significance in normal Portland cement clinkers, constituting 50–70% of the clinker content. Belite (C_2S) makes up around 15–30% of the clinker, aluminate (C_3A) accounts for 5–10% of the clinker, while ferrite (C_4AF) constitutes 5–15% of the clinker's composition [9]. After mixing with water, alite, belite, aluminate, and ferrite react with water to produce various hydration phases. The primary hydration products of alite and belite are portlandite ($Ca(OH)_2$) and calcium silicate hydrate (C-S-H). **Table 1** shows typical mineral compositions in cement.

The primary use of cement is to produce concrete. Concrete is a composite with cement, sand, aggregates (coarse bulks), and water. The combination of hydrated cement and aggregate gives concrete high mechanical strength, resulting in concrete becoming a widely used building material [11]. Cement itself is also widely used as a binding material. When mixed with water, cement can generate binding force and separate parts can be bonded together. Therefore, the steel is placed in concrete to increase the mechanical strength of material due to the binding force that cement can produce [12]. Another use of cement is for wellbore cementing, which is an important

Compound	Chemical formula	Oxide composition	Notation	Content (wt%)	Primary hydration products	Notes
Tricalcium silicate	Ca_3SiO_5	$(CaO)_3{\cdot}SiO_2$	C_3S	50–70	$Ca(OH)_2$ and C-S-H	Ensures the strength of cement at all stages of manufacture
Dicalcium silicate	Ca_2SiO_4	$(CaO)_2{\cdot}SiO_2$	C_2S	10–30	$Ca(OH)_2$ and C-S-H	Responsible for final strength, hydrates slowly
Tricalcium aluminate	$Ca_3Al_2O_6$	$(CaO)_3{\cdot}Al_2O_3$	C_3A	8–12	Various AFm (Al_2O_3-Fe_2O_3-mono) and AFt (Al_2O_3-Fe_2O_3-tri) phases	Hydrates rapidly, contributing to the strength of cement at an early stage; adding gypsum can control setting
Tetracalcium aluminoferrite	$Ca_4Al_2Fe_2O_{10}$	$(CaO)_4{\cdot}Al_2O_3{\cdot}Fe_2O_3$	C_4AF	8–12	Various AFm (Al_2O_3-Fe_2O_3-mono) and AFt (Al_2O_3-Fe_2O_3-tri) phases	Slow hydration; little effect on physical properties

Table 1.
Components in Class G cement formulation and their functionality [9, 10].

practice in oil and gas industry. Class G cement is the most popular wellbore cement with common applications in oil and gas wells, especially for high-pressure high temperature (HPHT) conditions and its compatibility with other additives. Class H cement is another commonly used oilwell cement. Compared with class G cement, class H cement has stronger retarding properties, making it ideal for cementing deep wells.

Cement and concrete are porous materials, and pores play a very important role in governing the water transport and strength properties of cement and concrete. The pore arrangement within cement-based materials comprises four distinct pore categories. These include: (a) gel pores, which are micro-sized pores spanning 0.5–10 nm; (b) capillary pores, representing mesopores with an average radius ranging from 5 to 5000 nm; (c) larger macropores resulting from deliberately entrained air; and (d) macropores arising from insufficient compaction [13]. In concrete, alongside the mentioned pores, cracks might develop at the interface between aggregate and mortar due to shrinkage. The gel pores, primarily sized at 1.5–2.0 nm, do not exert adverse effects on concrete strength through their porosity [13]; however, they are closely linked to creep and shrinkage. Conversely, capillary pores and other sizable pores notably contribute to strength reduction and diminished elasticity [14–16]. Therefore, when establishing an empirical correlation between concrete strength and porosity, the gel pores' role in overall porosity and pore size distribution can be disregarded without introducing notable errors. Consequently, for evaluating pore system characteristics affecting strength, the porosimetry test's maximum pressure should be adequate to trigger mercury intrusion into the tiniest capillary pore. In mercury porosimetry, a substantial portion of gel pores remains unintruded. Furthermore, sealed pores also remain unintruded. An additional limitation associated with mercury porosimetry is its focus on entry sizes rather than the authentic pore size linked to the ink bottle phenomenon [17]. Thus, the porosity calculated as mentioned earlier represents apparent porosity [13].

Different from cement, due to the presence of aggregates in concrete, the pore size distribution of concrete shows a complicated pattern. Concrete created using hydraulic cement binder can be considered a chemically bonded ceramic structure [13]. The cement's hydration reaction generates a composite of solids and a pore framework [18], meaning pores are inherent to concrete. Inadequate compaction can also lead to pore formation in concrete. This pore structure significantly governs concrete's essential attributes, particularly its strength [19, 20]. Well-compacted concrete containing dense, low-porosity aggregates can be envisioned as a multiphase substance, encompassing coarse aggregates within a mortar matrix. This matrix is composed of fine aggregates, solid cement hydrates, unhydrated cement, and the accompanying pore system [14]. However, the pore architecture in concrete's mortar differs notably from pores in well-compacted mortar formulated independently with the same ingredient proportions. This discrepancy arises from transition zone pores located at the mortar-aggregate junction [21, 22].

The capillary porosity of hardened cement paste hinges on the water-to-cement ratio, with this ratio also influencing transition zone porosity in concrete [14]. With the established relationship between strength and water-to-cement ratio, the concrete strength can be linked with pore structure characteristics [18, 23]. These relationships effectively guide concrete mix design. While some of these equations consider air content and concrete hydration, they often overlook structural pores stemming from inadequate compaction and other factors. Furthermore, the concrete's pore structure changes with hydration levels and chemical alterations due to aggressive environments [13].

Alternatively, a direct correlation could enable the estimation of *in situ* concrete strength based on the knowledge of its pore system features. However, for mix design purposes, the relationships between strength and water-to-cement ratio prove more practical. Key attributes of the pore system are porosity and pore size distribution, quantifiable through mercury intrusion porosimetry (MIP). Yet, MIP results are influenced by multiple factors, necessitating adjustments in the experimental approach [24–26]. Additionally, the smallest pore size that mercury can infiltrate depends on the maximum applied intrusion pressure. Thus, the porosity extent determinable by porosimetry rests on pore nature, the smallest anticipated pore size in the material, and the maximum intrusion pressure [25, 27].

2. CO_2-induced carbonation of cement and concrete

2.1 Carbonation of cement and concrete by low pressure CO_2

Under low pressure CO_2 (0.1 MPa CO_2 partial pressure or lower), the primary reaction of CO_2 with cement is that CO_2 reacts with Ca-bearing cement hydration products like calcium hydroxide and C-S-H to produce $CaCO_3$. Since the $CaCO_3$ production rate is usually higher than the consumption rates of calcium hydroxide and C-S-H, cement carbonation under low pressure CO_2 causes a decrease in porosity of cement, and this process is beneficial for the integrity of cement. The distribution and morphology of hydration products and carbonation products in cement before and after reaction with low pressure CO_2 obtained by high-magnification scanning electron microscopy (SEM) images are illustrated in **Figure 1**. $CaCO_3$ accumulates in the primary pores inside the cement and causes the closure of some pores and cracks, especially the small diameter pores [29]. Finally, the porosity of cement decreases after carbonation in low pressure CO_2 condition [30].

pH is the most important factor that governs the production of $CaCO_3$ and consumption of calcium hydroxide and C-S-H. The formation of a large amount of calcium hydroxide and C-S-H during hydration creates an alkaline environment in the interior of the cement and concrete, with a pH of approximately 11 [31]. When CO_2 is introduced, calcium hydroxide and C-S-H react with CO_2 and the pH is reduced. Under low pressure CO_2, the pH at the exterior of cement can drop to less than 9 [32], which is suitable for the precipitation of $CaCO_3$. $CaCO_3$ is stable under this pH, and the precipitation of $CaCO_3$ in cement pores reduces the permeability of cement, as well as increases the strength of cement. In short, the carbonization of cement will lead to many chemical and mechanical changes in the cement slurry, the most significant being changes in strength, porosity, pore size distribution, and chemical properties. In addition, it can cause shrinkage and cracking of the gelling matrix [33].

The reaction between concrete and CO_2 is almost the same as the reaction between cement and CO_2, since cement is the active component in concrete that reacts with CO_2. In general, only the cement experiences hydration during the curing process of concrete to produce hydration products like $Ca(OH)_2$ and C-S-H [34]. Another important component in concrete is aggregates (usually natural sands, which are fine aggregates and gravels, which are coarse aggregates) [35]. Aggregates do not react with CO_2. However, if steel is embedded in concrete and CO_2 is able to reach the steel in concrete, the steel may get corroded by CO_2, even when CO_2 presents in relatively low pressure.

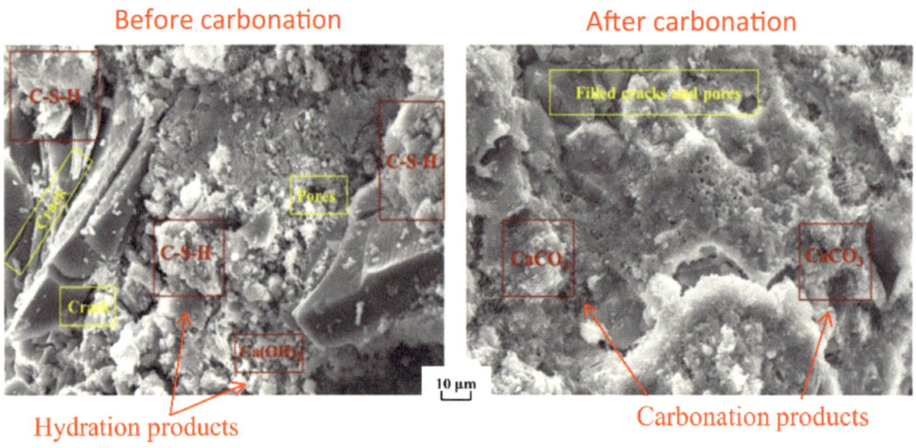

Figure 1.
Microstructure and morphology hydration products and carbonation products in cement before and after carbonation [28].

In general, cement hydration creates an alkaline environment inside concrete, which causes a passive film to form on the steel surface to protect steel from corrosion [36, 37]. The hydration products continue to react with CO_2 to form $CaCO_3$ to weaken the internal alkalinity of concrete with the continuous intrusion of CO_2. When the internal alkalinity of concrete is lost, the contact of CO_2 with the steel causes the corrosion of steel [38]. Under the influence of pore solution inside concrete, the steel reacts with CO_2 to produce iron carbonate ($FeCO_3$). With the process of reaction, $FeCO_3$ is converted into Fe(III)-bearing products like $Fe(OH)_3$ and Fe_2O_3 under the action of oxygen (O_2) and water (H_2O) [39]. Moreover, the corrosion products accumulate at the contact interface between steel and concrete, which causes the corrosion-induced cracks to occur owing to the lower density of corrosion products compared to the steel. The generation and expansion of induced cracks cause the separation of steel and concrete, which reduces the bonding strength of steel and concrete. Eventually, the structure of reinforced concrete is damaged and the strength of reinforced concrete is reduced under the influence of corrosion products, and the stability of reinforced concrete structure is affected by long-term influence of CO_2.

Though the global cement production process emits large amounts of CO_2, cement is able to absorb CO_2 back from the air, and thus cement is also an important carbon sink. From 1930 to 2013, the global cement industry emitted 38.1 billion tons of CO_2, while the carbon sink absorption by cement materials in the same period was as high as 16.5 billion tons, that is, 43% of the CO_2 emissions of cement industry processes during this period were absorbed back by cement materials after use [40, 41].

2.2 Carbonation of cement and concrete by high pressure and high concentration of CO_2

Under certain industrial practices like the exploitation of oil and gas from "sour" oil and gas reservoirs and geologic CO_2 utilization and storage, the pressure of CO_2 can be very high (e.g., 1–50 MPa). With the presence of water, the CO_2 becomes very corrosive to well cement and can cause degradation of well cement. Given this high-pressure CO_2, wellbore and its adjacent surroundings become chemically and mechanically unstable, posing a significant leakage risk to CO_2 storage systems. Long-term interaction between CO_2-saturated brine and wellbore cements, according to some experts, threatens wellbore integrity through calcium leaching [42]. During the degradation process of well cement, several geochemical reactions occur continuously because the cement is inherently unstable in aqueous environments rich in CO_2 [43]. In this case, the cement's sealing efficacy is determined by both environmental factors and the chemical composition and additives contained in it.

After exposure to carbonic acid under HPHT conditions, structural transformation, carbonation, and leaching are the primary mechanisms responsible for the physical and mechanical degradation of well cement. When C-S-H and $Ca(OH)_2$ react with carbonic acid to form $CaCO_3$, the process known as carbonation occurs. This process is quite similar to the carbonation process that occurs under low CO_2 pressure condition. A major difference between low CO_2 pressure condition and high CO_2 pressure condition is the corresponding aqueous phase pH. The aqueous phase pH can drop to <3 when the aqueous phase pH is in equilibrium with a high-pressure CO_2 phase (i.e., 1 MPa or higher). Under this low-pH condition, leaching occurs as the result of calcium ions migrating to the aggressive solution, which results in a progressive dissolution of cement paste [42, 44]. Significant cement leaching when exposed to carbonic acid under high CO_2 pressure conditions is not observed when cement is exposed to carbonic acid under low CO_2 pressure conditions (0.1 MPa CO_2 partial pressure or lower).

When wellbore cement is exposed to brine containing CO_2, the mechanical behavior of the cement is determined by the pace at which two competing mechanisms are at work: carbonation and leaching [45]. Carbonation drives the reaction front forward, whereas leaching occurs behind it. The fundamental reaction mechanisms are described as follows. The following chemical reactions occur when CO_2 dissolves in brine solution and attacks wellbore cement [46].

Step 1 is the formation of carbonic acid: As shown in Eq. (1), when CO_2 goes into a subsurface storage formation, it dissolves in the brine that's left over, generating carbonic acid (H_2CO_3). This results in a considerable decrease in pH [47].

$$CO_2 + H_2O \leftrightarrow H_2CO_3 \leftrightarrow H^+ + HCO_3^- \leftrightarrow 2H^+ + CO_3^{2-} \tag{1}$$

Step 2 is the carbonation of the cementitious compounds: Carbonic acid primarily interacts with portlandite ($Ca(OH)_2$) in the cement matrix, converting it into calcium carbonate and water.

i. Carbonation of calcium hydroxide.

$$Ca(OH)_2 + 2H^+ + CO_3^{2-} \rightarrow CaCO_3 + 2H_2O \tag{2}$$

This reaction, if not overreacted, benefits the wellbore's mechanical properties. Calcite, in particular, can increase the mechanical strength of the cement layer while decreasing porosity, resulting in a decrease in permeability, which will decrease the potential for CO_2 leakage [42].

ii. Carbonation of calcium silicate hydrate

a. Complete carbonation

$$3CaO{\cdot}2SiO_2{\cdot}3H_2O(s) + 3H_2CO_3(aq) \rightarrow 3CaCO_3(s) + 6H_2O + 2SiO_2(s) \tag{3}$$

b. Incomplete carbonation

$$3CaO{\cdot}2SiO_2{\cdot}3H_2O(s) + H_2CO_3(aq) \rightarrow C-S-C(s) \tag{4}$$

where C-S-C is the calcium silicate carbonate group.

The rapid carbonation of portlandite occurs at 107°C and up to 177°C with abundant presence of CO_2, precipitating $CaCO_3$ that may apply stress on surrounding cement, which results in a drop in mechanical strength. Therefore, moderate carbonation is beneficial for the cement's mechanical properties, while excessive carbonation impairs the mechanical strength of cement.

Step 3 is leaching out of calcium carbonates: The reaction between $CaCO_3$ and CO_2 created water-soluble $Ca(HCO_3)_2$ (Eq. (5)), and the continuous dissolution of $Ca(HCO_3)_2$ increased the cement's porosity throughout the later stages of carbonation.

$$CaCO_3(s) + H_2CO_3(aq) \rightarrow Ca(HCO_3)_2(aq) \tag{5}$$

As shown in **Figure 2**, the relationship between the inward diffusion of carbonated water and Ca^{2+} diffusion out of the cement is illustrated, along with the precipitation and dissolution of calcium carbonate. Normally, the consumption of portlandite occurs first and then the reaction involving C-S-H follows. Due to the presence of multiphasic components, many complex reactions may occur simultaneously, and the illustrated reactions in **Figure 2** are simplified reactions. A schematic of three distinctive layers as a result of CO_2 and wellbore cement interaction is shown in **Figure 2** and a computed tomography (CT) scanning image showing the three layers is provided in **Figure 3**.

The increase of CO_2 pressure and concentration raises the carbonation reaction rate of cement-based materials, which causes a large amount of $CaCO_3$ to form and accumulate in pores. In the initial stage of carbonation, the precipitation of $CaCO_3$ in pores makes the structure dense and reduces the porosity of cement-based materials, as shown in **Figure 4**. However, when $CaCO_3$ precipitation reaches the threshold value that the concrete can bear, the pores may be damaged by excessive $CaCO_3$ precipitation to produce microcracks [30]. Moreover, C-S-H decalcification may induce shrinkage during the carbonation reaction, and the shrinkage can also cause the generation of microcracks in cement-based materials [49]. When microcracks are generated inside cement-based materials, the material structure is destroyed and the penetration of CO_2 into cement-based materials becomes easy. Meanwhile, more hydration products come into contact with CO_2 to produce more $CaCO_3$. A large amount of $CaCO_3$

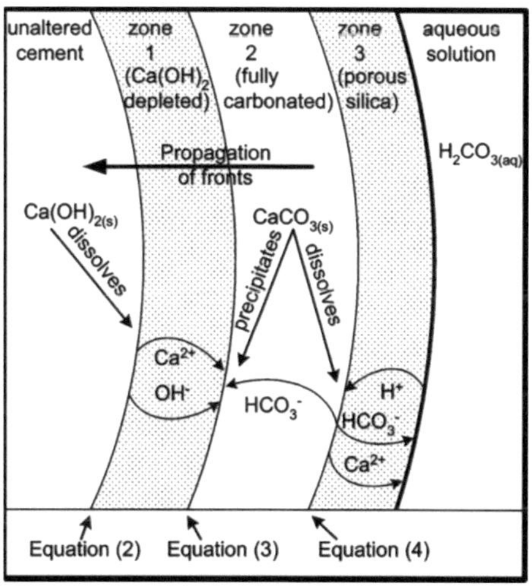

Figure 2.
The relationship between the inward diffusion of carbonated water and Ca^{2+} diffusion out of the cement, along with the precipitation and dissolution of calcium carbonate [48].

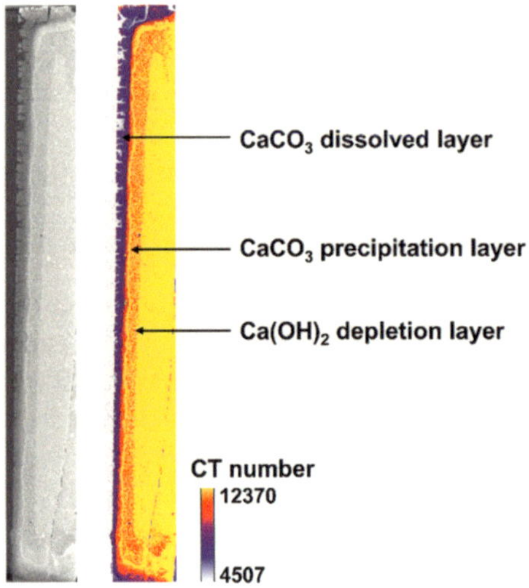

Figure 3.
Computed tomography (CT) scan image showing degradation of well cement and formation of three distinctive layers by CO_2 under geologic CO_2 sequestration conditions.

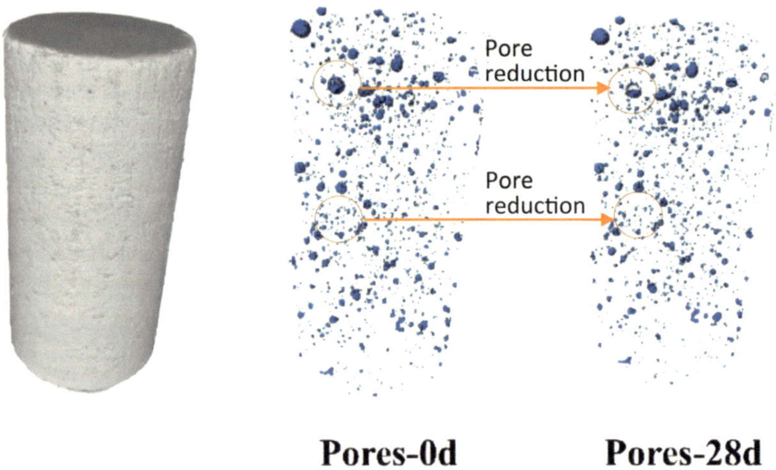

Figure 4.
Computed tomography (CT) scan image showing porosity reduction of cement-based materials by CO_2.

precipitation further promotes the expansion of microcracks. Finally, the structural failure and strength loss occur in cement-based materials caused by excessive carbonation [50, 51]. Under the influence of high pressure and high concentration of CO_2, cement-based materials are prone to excessive carbonation, which induces the porosity to decrease but weakens the strength of cement-based materials.

Wellbore cement's physical and mechanical properties are strongly influenced by temperature when used in a CO_2-rich environment. Previous research has demonstrated that Class G wellbore cement becomes stronger in mechanical strength while becoming less porous and permeable at 107°C [34]. When the carbon capture and storage (CCS) wellbores are permanently plugged, it is important to consider the durability of cement at elevated temperatures. However, when CO_2 is injected into a wellbore, the temperature significantly drops due to the Joule-Thomson effect, so the wellbore cement needs to withstand extreme temperature variation. This temperature drop is applicable when either continuous injection or cyclic intermittent injection procedures are followed.

The CO_2 leakage risk through wellbore cement has been considered in large-scale risk assessment models. Viswanathan [52] developed a hybrid system level model, CO_2-Predicting Engineered Natural Systems (CO_2-PENS), for the assessment of wellbore leakage at a geologic CO_2 sequestration site. Through the integration of information from laboratory experiments at the process level and field observations/ experiments at the process level, the model provides a science-based prediction approach. This CO_2-PENS model can not only be used to screen potential sequestration sites but can also be used as a tool for assessing individual sites' performance as detailed site-specific data become available. Another widely used wellbore CO_2 leakage risk evaluation code is WellRisk, which is based on drift flux method. The drift flux method was first developed by Zuber and Findlay [53], assuming that the flow rate of CO_2 in a two-phase flow is a function of the mixed flow rate and the "drift flow rate" (i.e., the flow rate at which CO_2 bubbles rise in a stationary liquid by buoyancy) in the two-phase flow [54].

WellRisk offers several advantages over CO_2-PENS. Notably, it eliminates the need for inputting the wellbore permeability, effectively resolving the challenge of obtaining accurate permeability coefficient data. Furthermore, the drift flux model adopted by WellRisk mimics the realistic migration characteristics of fluids within the wellbore, resulting in higher calculation accuracy compared to the permeability coefficient method adopted by CO_2-PENS.

3. New research directions

1. Combination of CT and nanoscale SEM to investigate the pore structure evolution of cement exposed to CO_2.

The size of the pores contained in cement is mostly between microns and nanometers, and the use of ordinary CTs with a resolution of microns to millimeters is not enough to directly characterize the pore structure of cement. In short, ordinary CTs can only provide the sample-scale trend of pore structure evolution in cement. Due to the limitation of CT scan resolution, the structural evolution of individual nanopores in cement after reaction with CO_2 cannot be characterized by ordinary CT. In order to show the pore structure evolution of wellbore cement after CO_2 corrosion from both sample scale and pore scale, it is necessary to combine CT scanning and high-resolution observation techniques (such as nanoscale SEM). Nanoscale scanning electron microscopy can reveal the structure of individual pores in a small area, which is very helpful in understanding the pore structure evolution of cement and concrete induced by carbonation.

2. Development of new cement additives to mitigate CO_2-induced cement degradation.

Wellbore cement is chemically unstable under CO_2-rich conditions because exposure of its hydration products to CO_2 can cause physicochemical changes that are harmful to the cement matrix. The physicochemical changes can finally lead to cement degradation and loss of cement integrity. In order to enhance the performance of wellbore cement in CO_2-rich environment, additives must be added to the cement to enhance the corrosion-resisting performance of wellbore cement. To improve the resistance of cement to CO_2 corrosion, researchers have developed a variety of additives that can be incorporated into wellbore cement, such as volcanic ash, fiber, self-healing additives, nanoparticles, etc., and the development of economically suitable additives with effective resistance to CO_2 corrosion is a promising research topic.

3. CO_2-accelerated curing of concrete

Cement-based materials react with CO_2 through cement hydration products such as calcium hydroxide to produce stable calcium carbonate. CO_2-accelerated curing of concrete refers to the introduction of CO_2 to the production process of concrete, which can reduce CO_2 emissions and enhance the strength of concrete, mainly because the addition of CO_2 can produce more calcium carbonate during the cement hardening process, thereby increasing the strength of cement. Therefore, CO_2-accelerated curing of concrete is also an area that researchers are currently focusing on.

4. Summary

This chapter introduces the development history of cement and concrete, and analyzes the evolution of pore structure of cement and concrete by CO_2 carbonation. The results show that under low pressure CO_2, cement carbonation leads to the formation of a dense $CaCO_3$-rich area, resulting in reduced cement permeability and increased cement strength. For wellbore cement exposed to high pressure and high concentration of CO_2, a low-porosity calcite deposition layer is formed in the cement, and two highly porous layers due to the dissolution of calcite and hydration products are formed on both sides of the calcite deposition layer. For concrete exposed to atmospheric pressure CO_2, carbonation causes hydration products to form $CaCO_3$ and precipitate in the pores, reducing pore volume and total porosity. For concrete exposed to high-pressure CO_2, excessive accumulation of $CaCO_3$ eventually leads to pore expansion and cracking, resulting in a decrease in compressive strength after the compressive strength reaches its peak. The chapter also presents recent topics of interest, including the characterization of the nano- and microscopic pore structure of cement exposed to CO_2 using nanoscale SEM scanning combined with microcomputed tomography (micro-CT), the development of high-performance cement additives to resist CO_2 corrosion, and CO_2-accelerated curing of concrete. In summary, an in-depth understanding of the pore structure changes after carbonation of cement and concrete is important for advancing research on cement-based materials, so as to build safe, green, and sustainable infrastructure.

Acknowledgements

The authors are grateful for the funding support provided by the National Natural Science Foundation of China (Grant No. 42172315) and the Science and Technology Plan Project of Sichuan Province (2022YFSY0018).

Conflict of interest

The authors declare no conflict of interest.

Author details

Liwei Zhang*, Manguang Gan and Quan Xue
State Key Laboratory of Geomechanics and Geotechnical Engineering, Institute of Rock and Soil Mechanics, Chinese Academy of Sciences, Wuhan, Hubei, China

*Address all correspondence to: lwzhang@whrsm.ac.cn

References

[1] Cuezva S, Benavente D, Ivars J, Galan JM. Composition, uses, provenance and stability of rocks and ancient mortars in a Theban Tomb in Luxor (Egypt). Materials and Structures. 2016;**49**:941-960. DOI: 10.1617/s11527-015-0550-5

[2] Palomo A, Monteiro P, Martauz P, Bilek V, Fernandez-jimenez A, De Ciencias I, et al. Hybrid binders: A journey from the past to a sustainable future (opus caementicium futurum). Cement and Concrete Research. 2019;**124**:105829. DOI: 10.1016/j.cemconres.2019.105829

[3] Science Museum. Building the Modern World: Concrete and Our Environment [Internet]. 2021. Available from: https://www.sciencemuseum.org.uk/objects-and-stories/everyday-wonders/building-modern-world-concrete-and-our-environment

[4] Britannica. History of Cement [Internet]. 2023. Available from: https://www.britannica.com/technology/cement-building-material/History-of-cement

[5] Ashby MF. Materials and Sustainable Development. 2nd ed. Butterworth Heinemann: Oxford; 2023. 565 p. DOI: 10.1016/C2021-0-00557-5

[6] Han B, Ding S, Wang J, Ou J. Nano-Engineered Cementitious Composites: Principles and Practices. Singapore: Springer; 2018. 731 p. DOI: 10.1007/978-981-13-7078-6

[7] Yousuf N, Olayiwola O, Guo B, Liu N. A comprehensive review on the loss of wellbore integrity due to cement failure and available remedial methods. Journal of Petroleum Science and Engineering. 2021;**207**:109123. DOI: 10.1016/j.petrol.2021.109123

[8] Civil Engineering. Uses of Cement [Internet]. 2023. Available from: https://civiltoday.com/civil-engineering-materials/cement/46-uses-of-cement#:~:text=It is used in mortar for plastering%2C masonry,roofs and constructing lintels%2C beams%2C stairs%2C pillars%2C etc

[9] Taylor HFW. Cement Chemistry. 2nd ed. London: Thomas Telford Publishing; 1997. 439 p. DOI: 10.1680/cc.25929

[10] Mangadlao JD, Cao P, Advincula RC. Smart cements and cement additives for oil and gas operations. Journal of Petroleum Science and Engineering. 2015;**129**:63-76. DOI: 10.1016/j.petrol.2015.02.009

[11] Newell P. An ancient battle between environment and concrete. Nature Reviews Chemistry. 2021;**5**:513-514. DOI: 10.1038/s41570-021-00315-z

[12] Choi Y, Yuan RL. Experimental relationship between splitting tensile strength and compressive strength of GFRC and PFRC. Cement and Concrete Research. 2005;**35**:1587-1591. DOI: 10.1016/j.cemconres.2004.09.010

[13] Kumar R, Bhattacharjee B. Porosity, pore size distribution and in situ strength of concrete. Cement and Concrete Research. 2003;**33**:155-164. DOI: 10.1016/S0008-8846(02)00942-0

[14] Soroka I. Portland Cement Paste and Concrete. London: Macmillan; 1979. 338 p. DOI: 10.1007/978-1-349-03994-4

[15] Neville AM, Brooks JJ. Concrete Technology. Longman: Singapore; 1990. 464 p. isbn/0582988594

[16] Brandt AM. Cement Based Composites: Materials, Mechanical Properties and Performance. London: E&FN Spon; 1995. 544 p. DOI: 10.1201/9781482265866

[17] Diamond S. Mercury porosimetry: An inappropriate method for the measurement of pore size distributions in cement-based materials. Cement and Concrete Research. 2000;**30**:1517-1525. DOI: 10.1016/S0008-8846(00)00370-7

[18] Nagaraj TS, Banu Z. Generalization of Abrams' laws. Cement and Concrete Research. 1996;**26**:933-942. DOI: 10.1016/0008-8846(96)00065-8

[19] Därr GM, Ludwig U. Determination of permeable porosity. Matériaux et Construction. 1973;**6**:185-190. DOI: 10.1007/BF02479032

[20] Rostásy FS, Weiβ R, Wiedemann G. Changes of pore structure of cement mortars due to temperature. Cement and Concrete Research. 1980;**10**:157-164. DOI: 10.1016/0008-8846(80)90072-1

[21] Winslow D, Liu D, Lafayette W. The pore structure of paste in concrete. Cement and Concrete Research. 1990;**20**:227-235. DOI: 10.1016/0008-8846(90)90075-9

[22] Aguado A, Gettu R, Shah S. Concrete Technology: New Trends, Industrial Applications. London: E&FN Spon; 1994. 366 p. DOI: 10.1201/9781482271584

[23] Popovics S. New formulas for the prediction of the effect of porosity on concrete strength. ACI Journal Proceedings. 1981;**82**:136-146. DOI: 10.14359/10321

[24] Laskar MAI, Kumar R, Bhattacharjee B. Some aspects of evaluation of concrete through mercury intrusion porosimetry. Cement and Concrete Research. 1997;**27**:93-105. DOI: 10.1016/s0008-8846(96)00192-5

[25] Winslow DN, Diamond S. A mercury porosimetry study of the evolution of porosity in Portland cement: Technical publication. ACI Journal Proceedings. 1970;**5**:564-585. DOI: 10.5703/1288284314510

[26] Diamond S. A critical comparison of mercury porosimetry and capillary condensation pore size distributions of Portland cement paste. Cement and Concrete Research. 1972;**2**:148-151. DOI: 10.1016/0008-8846(72)90032-4

[27] Kumar R. Strength and permeation quality of concrete through mercury intrusion porosimetry [thesis]. Delhi, India: Indian Institute of Technology Delhi; 1997. Available from: https://libcat.iitd.ac.in/cgi-bin/koha/opac-detail.pl?biblionumber=163559

[28] Liang C, Lu N, Ma H, Ma Z, Duan Z. Carbonation behavior of recycled concrete with CO_2-curing recycled aggregate under various environments. Journal of CO_2 Utilization. 2020;**39**:101185. DOI: 10.1016/j.jcou.2020.101185

[29] Ghahari SA, Ramezanianpour AM, Ramezanianpour AA, Esmaeili M. An accelerated test method of simultaneous carbonation and chloride ion ingress: Durability of silica fume concrete in severe environments. Advances in Materials Science and Engineering. 2016;**2016**:1-12. DOI: 10.1155/2016/1650979. Available from: https://www.hindawi.com/journals/amse/2016/1650979/

[30] Zhao K, Liang Y, Ji T, Lu Y, Lin X. Effect of activator types and concentration of CO_2 on the steel corrosion in the carbonated alkali-activated slag concrete.

Construction and Building Materials. 2020;**262**:120044. DOI: 10.1016/j.conbuildmat.2020.120044

[31] Ho DWS, Lewis RK. Carbonation of concrete and its prediction. Cement and Concrete Research. 1987;**17**:489-504. DOI: 10.1016/0008-8846(87)90012-3

[32] Papadakis VG, Fardis MN, Vayenas CG. Effect of composition, environmental factors and cement-lime mortar coating on concrete carbonation. Materials and Structures. 1992;**25**:293-304. DOI: 10.1007/bf02472670

[33] Šavija B, Lukovic M. Carbonation of cement paste: Understanding, challenges, and opportunities. Construction and Building Materials. 2016;**117**:285-301. DOI: 10.1016/j.conbuildmat.2016.04.138

[34] Wang J, Niu D, Zhang Y. Microstructure and mechanical properties of accelerated sprayed concrete. Materials and Structures. 2016;**49**:1469-1484. DOI: 10.1617/s11527-015-0589-3

[35] Ulm FJ, Constantinides G, Heukamp FH. Is concrete a poromechanics material?—A multiscale investigation of poroelastic properties. Materials and Structures. 2004;**37**:43-58. DOI: 10.1007/bf02481626

[36] Brunet JPL, Li L, Karpyn ZT, Huerta NJ. Fracture opening or self-sealing: Critical residence time as a unifying parameter for cement–CO_2–brine interactions. International Journal of Greenhouse Gas Control. 2016;**47**:25-37. DOI: 10.1016/j.ijggc.2016.01.024

[37] Silva A, Neves R, De Brito J. Statistical modelling of carbonation in reinforced concrete. Cement and Concrete Composites. 2014;**50**:73-81. DOI: 10.1016/j.cemconcomp.2013.12.001

[38] Liang C, Pan B, Ma Z, He Z, Duan Z. Utilization of CO_2 curing to enhance the properties of recycled aggregate and prepared concrete: A review. Cement and Concrete Composites. 2020;**105**:103446. DOI: 10.1016/j.cemconcomp.2019.103446

[39] De Motte R, Basilico E, Mingant R, Kittel J, Ropital F, Combrade P, et al. A study by electrochemical impedance spectroscopy and surface analysis of corrosion product layers formed during CO_2 corrosion of low alloy steel. Corrosion Science. 2020;**172**:108666. DOI: 10.1016/j.corsci.2020.108666

[40] Xi F, Davis SJ, Ciais P, Crawford-brown D, Guan D, Pade C, et al. Substantial global carbon uptake by cement carbonation. Nature Geoscience. 2016;**9**:880-883. DOI: 10.1038/NGEO2840

[41] Mora P, Sanjuan MA. Sequestration of CO_2 by concrete carbonation. Environmental Science & Technology. 2010;**44**:3181-3186. DOI: 10.1021/es903581d

[42] Omosebi O, Maheshwari H, Ahmed R, Shah S, Osisanya S, Santra A, et al. Investigating temperature effect on degradation of well cement in HPHT carbonic acid environment. Journal of Natural Gas Science and Engineering. 2015;**26**:1344-1362. DOI: 10.1016/j.jngse.2015.08.018

[43] Montegrossi G, Huet B, Vaselli O, Hern A, Virgili G. A study of wellbore cement alteration controlled by CO_2 leakage in a natural analogue for geological CO_2 storage. Applied Geochemistry. 2017;**86**:13-25. DOI: 10.1016/j.apgeochem.2017.09.010

[44] Zhang J, Peng Z, Zou C, Chen D. Improving the carbonation resistance of cement stone for oil wells by a polymer with acid response characteristic. Journal

of Petroleum Science and Engineering. 2018;**164**:382-389. DOI: 10.1016/j.petrol.2018.01.080

[45] Duguid A, Scherer GW. Degradation of oilwell cement due to exposure to carbonated brine. International Journal of Greenhouse Gas Control. 2010;**4**:546-560. DOI: 10.1016/j.ijggc.2009.11.001

[46] Marshdi QSR. Benefits of using mineral additives, as components of the modern oil-well cement. Case Studies in Construction Materials. 2018;**8**:455-458. DOI: 10.1016/j.cscm.2018.03.010

[47] Abid K, Gholami R, Choate P, Hari B. A review on cement degradation under CO_2-rich environment of sequestration projects. Journal of Natural Gas Science and Engineering. 2015;**27**:1149-1157. DOI: 10.1016/j.jngse.2015.09.061

[48] Kutchko BG, Strazisar BR, Dzombak GV, Lowry DA, Thaulow N. Degradation of well cement by CO_2 under geologic sequestration conditions. Environmental Science & Technology. 2007;**41**:4787-4792. DOI: 10.1021/es062828c

[49] Auroy M, Poyet S, Le P, Torrenti J, Charpentier T, Moskura M, et al. Impact of carbonation on unsaturated water transport properties of cement-based materials. Cement and Concrete Research. 2015;**74**:44-58. DOI: 10.1016/j.cemconres.2015.04.002

[50] Xue Q, Zhang L, Mei K, Wang L, Wang Y, Li X, et al. Evolution of structural and mechanical properties of concrete exposed to high concentration CO_2. Construction and Building Materials. 2022;**343**:128077. DOI: 10.1016/j.conbuildmat.2022.128077

[51] Xue Q, Zhang L, Mei K, Li X, Newell P, Wang Y, et al. CO_2-induced evolution of chemical, structural and mechanical properties of reinforced concrete: A review. Construction and Building Materials. 2022;**353**:129069. DOI: 10.1016/j.conbuildmat.2022.129069

[52] Stauffer PH, Kaszuba JP, Carey JW, Olsen SC, Keating GN, Kavetski D, et al. Development of a hybrid process and system model for the assessment of wellbore leakage at a geologic CO_2 sequestration site. Environmental Science & Technology. 2008;**42**:7280-7286. DOI: 10.1021/es800417x

[53] Zuber N, Findlay JA. Average volumetric concentration in two-phase flow systems. Journal of Heat Transfer. 1965;**87**:453. DOI: 10.1115/1.3689137

[54] Pan L, Oldenburg CM, Wu Y, Pruess K. Wellbore flow model for carbon dioxide and brine. Energy Procedia. 2009;**1**:71-78. DOI: 10.1016/j.egypro.2009.01.012

Chapter 6

Application of Porous Media in Heat-Transfer and Mass-Transfer Enhancements: A Brief Review

Guojun Yu and Huijin Xu

Abstract

The heat and/or mass transfer is crucial in various energy conversion and storage systems such as heat exchangers and energy storage systems, since they highly affect the efficiency of energy conversion and transport. Enhancing the heat and/or mass transfer within these systems is the most important means to improve system efficiency. Porous media have found wide application in enhancing the heat conduction, mass diffusion, or both, for different energy conversion and storage systems. In this chapter, a brief review on the application of different porous media for transport enhancement in various systems was made, indicating that using porous media is capable of enhancing the transport ability appreciably, sometimes being up to hundreds of times in some physical problems. This review could provide some insight into the transport enhancement design of various energy conversion and storage systems, which is especially important in the background of carbon neutralization.

Keywords: porous media, heat-transfer enhancement, mass-transfer enhancement, energy conversion, energy storage

1. Introduction

Porous media structures are widely existing around us, like soil, rocks, wood, and many other solid objects. Some features of these porous structures, like large surface area, low-density, permeability, selectivity, adsorptivity, and thermal conductivity, have inspired us to explore their potential applications. Some artificial porous media with specified porous structure, like activated carbon, zeolites, and sponges, metal foams, have been designed for various purposes. These natural and synthetic porous materials have received a wide range of applications, including filtration and separation, chemical reaction, adsorption, heat exchanger, hydrogen storage, fuel cells, and so forth.

This chapter, however, will focus on the applications of the porous media in heat-transfer and mass-transfer enhancements. The enhancements in heat and mass transfer have been attracting significant attention, especially in the background of carbon neutralization, since they are capable of enhancing the energy efficiency and thus reducing the carbon emission. The unique characteristics of porous media, such as large surface area, rich interconnected pore structure, and high thermal conductivity, make them good candidates to enhance the heat and mass transfer in various energy systems.

Therefore, this chapter will make a brief review on where and how the porous media have been employed to enhance the heat and mass transfer. To the authors' knowledge, there is no systematic review on the application of porous media in terms of heat-transfer and mass-transfer enhancements. So, we hope this chapter would provide a reference to the researchers or engineers when they are developing innovative solutions to enhance the heat and/or mass transfer so as to enhance the efficiency and performance of various industrial processes.

2. Application of porous media in heat-transfer enhancement

Enhancement in heat transfer by porous media can be realized from increase in either thermal conductivity, convective heat transfer, or radiation heat transfer. However, the radiation is not discussed in this chapter due to space limit. Therefore, in this part, some applications of porous media to enhance the heat transfer will be discussed from the heat conduction and convection aspects.

2.1 Heat conduction enhancement by porous media

The solid matrix of some porous media has larger thermal conductivity than the working medium (e.g., phase change materials), so the working medium can be embedded in some porous media for the enhancement in thermal conductivity and thus the thermal performance. The applications with this regard mainly associated with latent-heat energy storage systems.

Phase change materials (PCMs) are usually of low thermal conductivity, which seriously restrains the energy storage efficiency. Since the solid matrix of porous media bears large thermal conductivity, the PCMs can be embedded in the porous matrix, and therefore, the high thermal conductivity of porous media enables efficient heat transfer between the PCMs and the heat exchangers. The porous media used for this application are mainly divided into two groups, graphite-based porous media and metal foams.

The widely used graphite-based porous media are mainly divided into two groups: graphite foam and expanded graphite (EG). Graphite foam is usually produced by a foaming process that creates interconnected pores within the graphite material, thus creating a highly porous structure. However, expanded graphite is produced by treating the natural graphite with chemicals or high temperatures. This treatment causes the layers of graphite to separate and expand, creating a porous structure with a high surface area. The resulting expanded graphite can be used to embed PCMs to produce composite PCMs with large conductivity.

Zhong et al. [1] prepared four composite PCMs by paraffin wax and mesophase pitch-based graphite foams (GFs) with different thermal properties and pore sizes. The polarizing optical microscopy (POM) pictures of the polished surface of the four composite PCM samples are shown in **Figure 1**. After the test, they found that the thermal diffusivity of the paraffin-GFs can be enhanced by 190, 270, 500, and 570 times, respectively, as compared with that of pure paraffin wax. Zhong et al. [2] synthesized three kinds of porous heterogeneous composite PCMs with EG and binary molten salts ($LiNO_3$-KCl, $LiNO_3$-$NaNO_3$ and $LiNO_3$-NaCl), as shown in **Figure 2**. They found that the heat conductivity of all binary molten salts was enhanced by 4.9–6.9 times after impregnation with EG. Sari et al. [3] prepared different paraffin/EG composites with

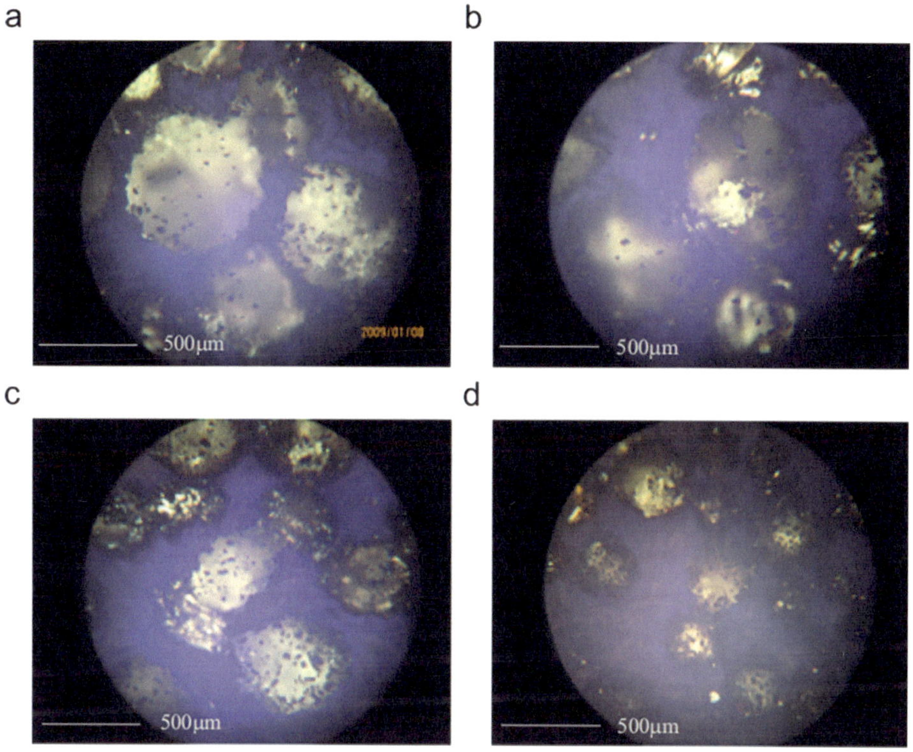

Figure 1.
POM pictures of the polished surface of the four filled foam samples [1].

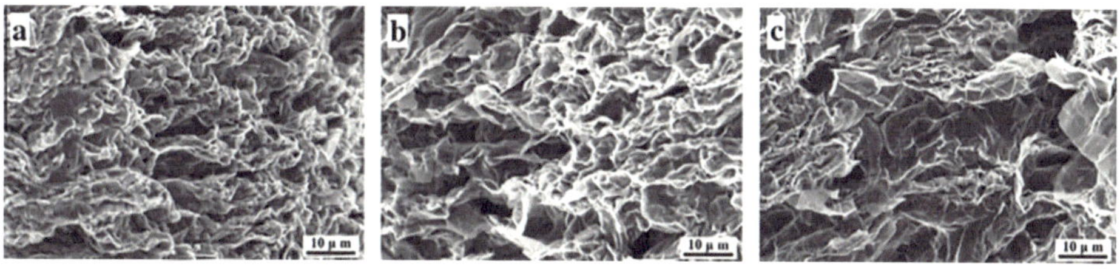

Figure 2.
SEM photographs of porous heterogeneous composite phase change materials: (a) $LiNO_3$-KCl/EG, (b) $LiNO_3$-$NaNO_3$/EG, and (c) $LiNO_3$-NaCl/EG [2].

the EG mass fraction of 2, 4, 7, and 10%. Heat conductivities of the pure paraffin and the composite PCMs were measured as 0.22, 0.40, 0.52, 0.68, and 0.82 W/(mK), respectively, indicating that the heat conductivity of the paraffin/EG composite was enhanced by approximately 2–4 times compared to that of pure paraffin.

Another porous media usually used to enhance the heat conductivity of PCMs are metal foams, which are typically produced through a process called foaming, where a mixture of metal powder, a blowing agent, and a binder are heated or chemically treated to create a foam-like structure. Metal foams find wide applications in latent heat thermal energy storage with PCMs.

Xiao et al. [4] prepared the paraffin/nickel foam and paraffin/copper foam PCMs, each of which was designed with three different pore sizes (5PPI, 10PPI, and 25PPI), as shown in **Figure 3**. Their results indicated that the heat conductivity of the paraffin/nickel foam composite was nearly three times larger than that of pure paraffin, while the paraffin/copper foam composite was 15 times larger. Fleming et al. [5] investigated the heat-transfer enhancement of a shell-and-tube latent heat thermal storage unit, as shown in **Figure 4**, by the addition of open-cell aluminum foam. Their results showed that the inclusion of aluminum foam significantly enhances the heat transfer coefficient by 100% for the melting process and by 20% for the

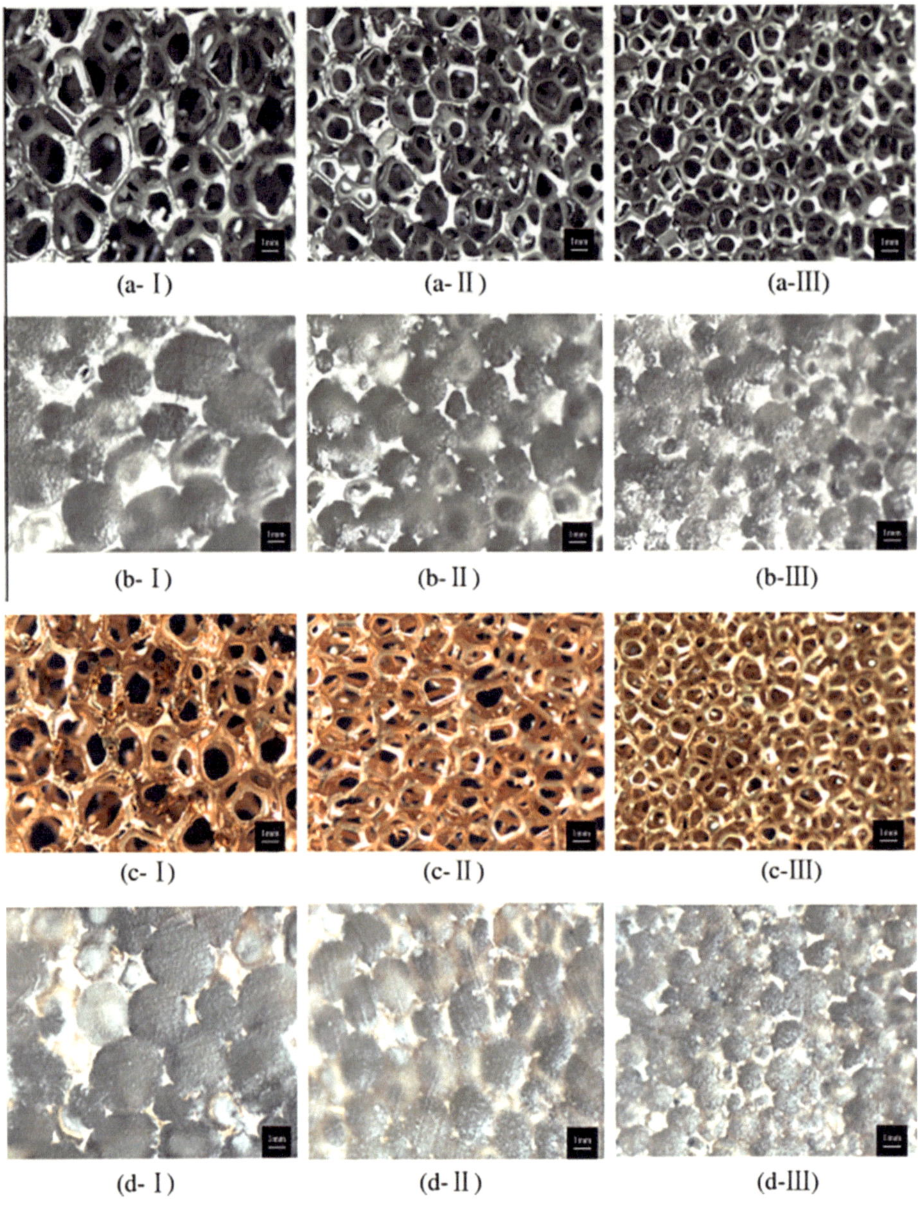

Figure 3.
Images of metal foam and paraffin/metal foam composite PCMs with different pore sizes: (a) nickel foams, (b) paraffin/nickel foam composite PCMs, (c) copper foams, and (d) paraffin/copper foam composite PCMs (I: 5PPI, II: 10PPI, III: 25PPI) [4].

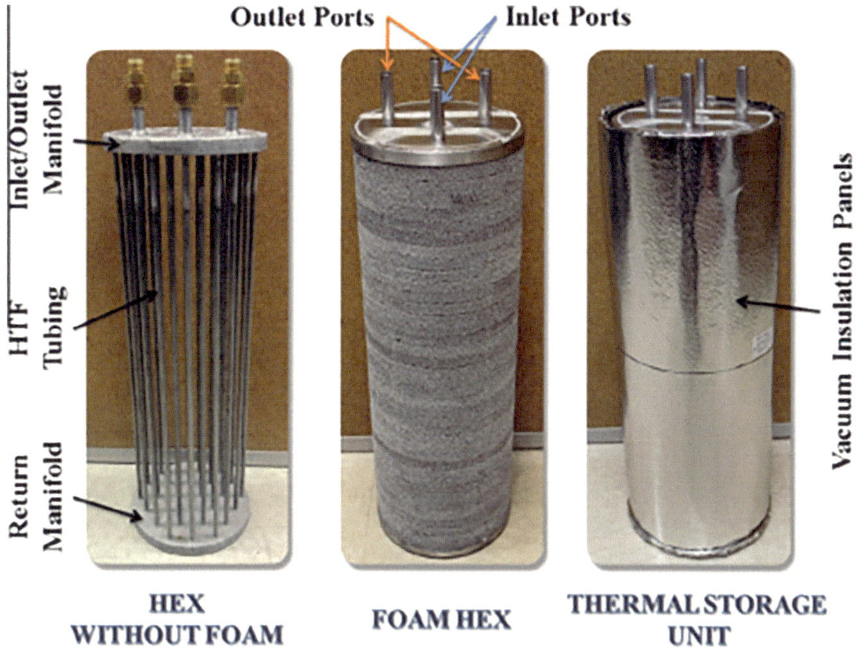

Figure 4.
The dual-pass heat exchangers with and without foam and the complete storage unit (without lid) insulated with vacuum insulation panels [5].

solidification process. It was also found that the natural convection during melting was suppressed by the foam, but the suppression was still outweighed by the thermal conductivity enhancement.

Some other research regarding the application of porous materials/foams in PCMs for the heat-transfer enchantment purpose can be referred to Ref. [6].

2.2 Convection heat-transfer enhancement by porous media

The enhancement in convection heat transfer by porous media attributes to a few factors: Firstly, the porous media have larger specific surface area, providing more contact points between liquids and solids and thus enhancing the convection heat transfer; secondly, the interconnected pores in porous media allow for more efficient transport of fluids through the material than solids, leading to heat-transfer enhancement by convection, which carries heat away from the hot regions and distributes it throughout the material; thirdly, the porous structure at the wall surface in a flow stream causes some micro local flow, which enhances the heat transfer coefficient; and fourthly, the porous structure provides more nucleation sites for boiling, and enhances the surface wettability and thus promotes the formation of thin liquid films for condensation, both leading to increased heat transfer coefficients; the abovementioned four aspects of porous media collectively make them effective in improving heat transfer performance.

The metal foams find good application in tubes to enhance the convection heat transfer. These metal foam embedded tubes are widely used in various heat exchangers due to high heat transfer efficiency. Huang et al. [7] inserted a porous media in the core of a tube for the heat-transfer enhancement purpose, as shown in **Figure 5**.

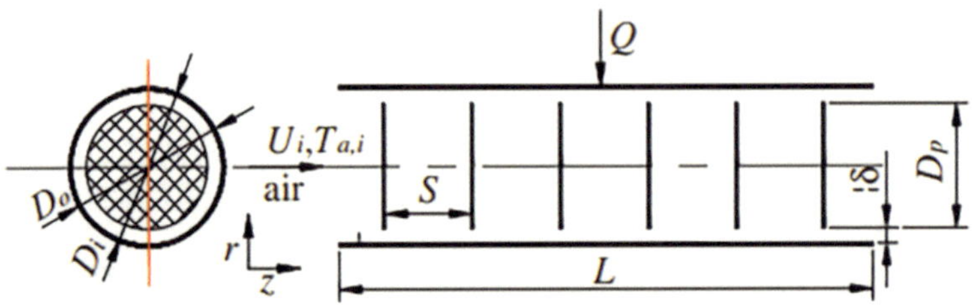

Figure 5.
Schematic diagram of the problem [7].

Their results showed that the heat transfer rate in the tube with inserted porous media whose diameter approaches the diameter of the tube is about 1.6–5.5 times larger than those of the smooth tube cases in laminar, transitional, and turbulent flows. Ming et al. [8] studied the similar problem, finding that inserting the metal foam in the tube brings a notable increase of Nusselt number and that good heat transfer performance was achieved if the diameter of the metals foam approaches the tube diameter.

Metal foams are also used in other geometries to enhance the heat transfer. For example, Vazifeshenas et al. [9] carried out both numerical and experimental research to assess the thermo-hydraulic behavior of open aluminum metal foam sheet adjacent to the heated wall in role of polymer electrolyte membrane (PEM) fuel cell end plate shown in **Figure 6**. The result showed that open aluminum metal foam is good for heat transfer and that the heat transfer rate is positively proportional to the porosity of the foam.

Another application of porous media for heat-transfer enhancement is associated with boiling. Bai et al. [10] investigated the enhancement of boiling of anhydrous ethanol in the porous-coated microchannels. In their research, three porous-coated microchannels were prepared by fabricating metallic porous coating with copper particles of different diameters in the bottom of the microchannels by a solid-state

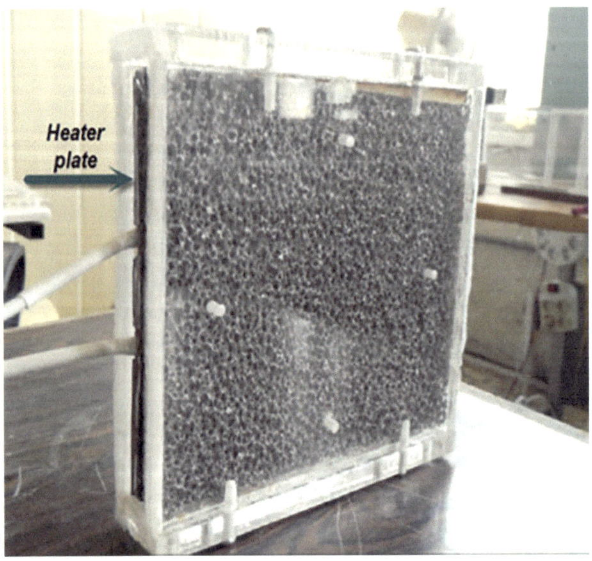

Figure 6.
Single-cell metal foam adjacent to electrical heater plate [9].

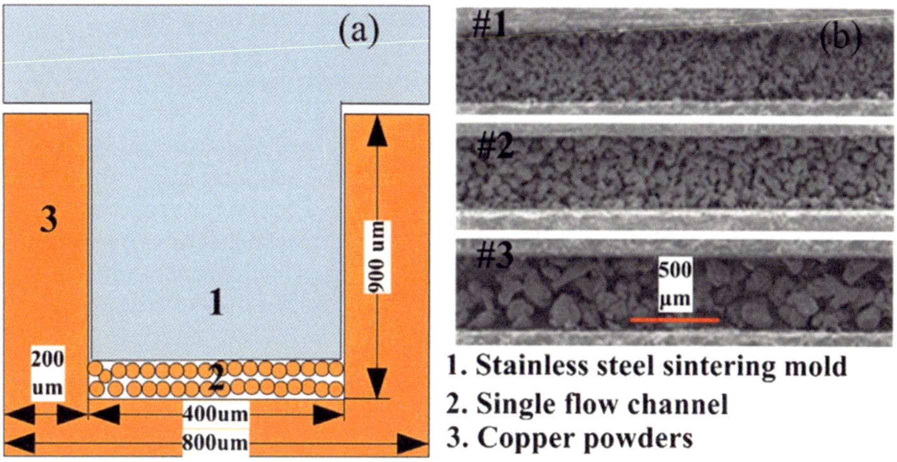

Figure 7.
(a) Schematic of solid-state sintering process; (b) SEM images of porous-coated microchannels (#1, #2, and #3) [10].

sintering technology, as shown in **Figure 7**. As indicated by their results, a dramatic enhancement of flow boiling heat transfer in the porous-coated microchannels was obtained, due to the preferable bubble nucleation condition of porous coatings. Ma et al. [11] investigated the pool boiling heat transfer on the surface of different kinds of multilayer gradient aperture porous copper they prepared, as shown in **Figure 8**. They found that bubble nucleation on the porous surface emerges at much lower heat flux (50% lower) than that of the smooth surface.

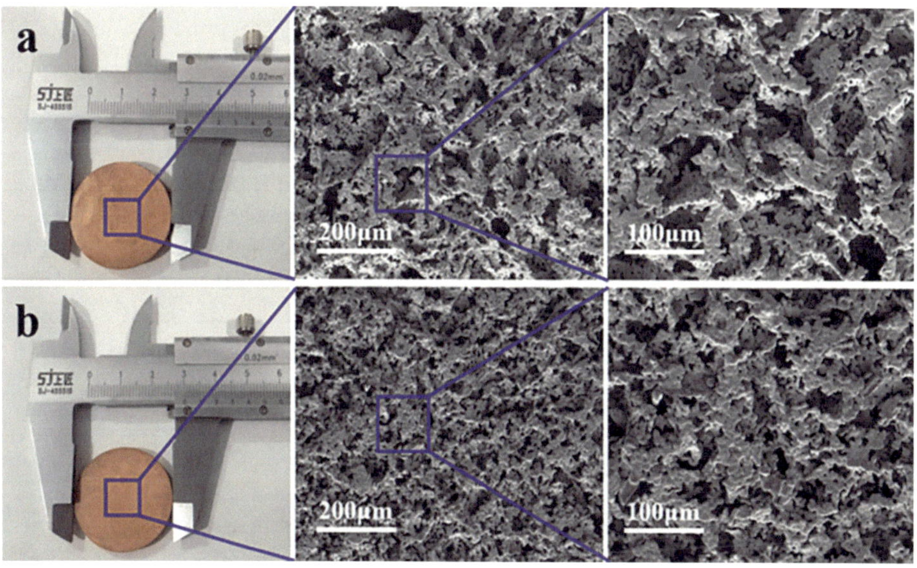

Figure 8.
Macro-morphology and SEM photos of the top and bottom surfaces of the 4-layer sample: (a) top view where the pore is 100 mesh; (b) bottom view where the pore is 300 mesh [11].

3. Application of porous media in mass-transfer enhancement

The mass-transfer enhancement by porous media attributes to the porous interconnected voids within their structure, which can be of various sizes and shapes and allow for the stronger flow of fluids than that through the low-permeability solids.

The most well-known application of porous media in mass-transfer enhancement is in crude oil and natural gas production industry. Sometimes, the crude oil reservoir or natural gas reservoir is of low permeability; some artificial fractures have to be constructed by a technique called hydraulic fracturing, which forms an interconnected porous zone, as shown in **Figure 9**, and enhances the flow of crude oil or natural gas in the reservoir. Geometries of various fractures can be seen in **Figure 10**.

Another typical example of the application of porous media in mass-transfer enhancement, actually both heat- and mass-transfer enhancements, is in the exploitation of geothermal energy from dry rocks. There is a huge amount of heat energy stored in the dry rock, but the dry rock is of pretty low permeability, through which the fluid hardly flows to extract heat energy. An enhanced geothermal system (EGS) [14], as shown in **Figure 11**, was proposed to exploit the thermal energy in these dry rocks, by drilling wells into hot rock and fracturing the rock sufficiently to enable the water to flow between the wells. The fluid flows along permeable pathways, picking up heat, and exits the reservoir through production wells. The fractures allow for the circulation of the fluid through the heat reservoir, increasing the contact area between the fluid and the hot rock. This enhances the mass transfer as well as heat transfer from the rock to the fluid. The characteristics of fractures have a great impact on the heat transfer performance, details of which can be found in Refs. [15, 16].

Mass-transfer enhancement by porous media also finds application in hydrogen storage. Among various hydrogen storage methods, hydrogen storage in metal hydride (e.g., magnesium hydride) is a very promising technique due to large storage capacity [17, 18]. However, one of the most important factors influencing the storage efficiency is the transport of hydrogen in the metals that are of pretty low-permeability, because the hydrogen has to transport within the metal and chemically react with the metals in the bulk of the material. To solve this problem, the metal powder, which behaves like a porous media, is used instead of solid metal. The widely used hydrogen storage system based on metal powder is shown in **Figure 12**. The hydrogen can transport through the metal powder and chemically react with it, and the produced hydride is still in a powder form, which also allows the hydrogen to flow through it. The porous structure formed by the metal powder effectively enhances the mass transport of the hydrogen to ensure the chemical reaction. A real application of this metal powder

Figure 9.
Intersection of hydraulic fractures with natural horizontal fractures in multistage horizontal well [12].

Figure 10.
Different fractures of the reservoir [13].

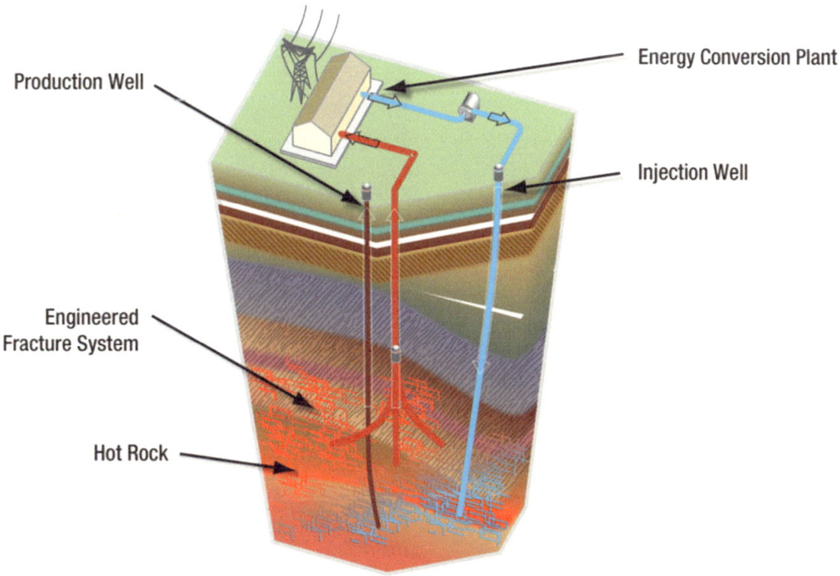

Figure 11.
EGS cutaway diagram [14].

(Mg) for hydrogen storage is shown in **Figure 13**. As can be seen, the stack of Mg/ MgH_2 presents obvious porous structure.

In addition, porous media can also be used to enhance the mass transfer in electrode of fuel cells or other batteries. These porous electrodes can provide a larger

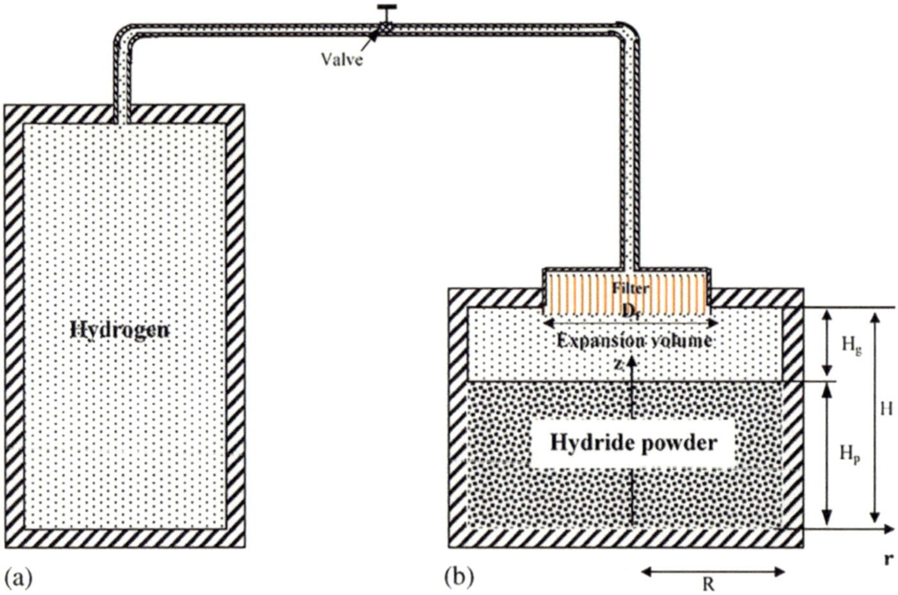

Figure 12.
Geometrical configuration of (a) hydrogen tank and (b) the reactor [19].

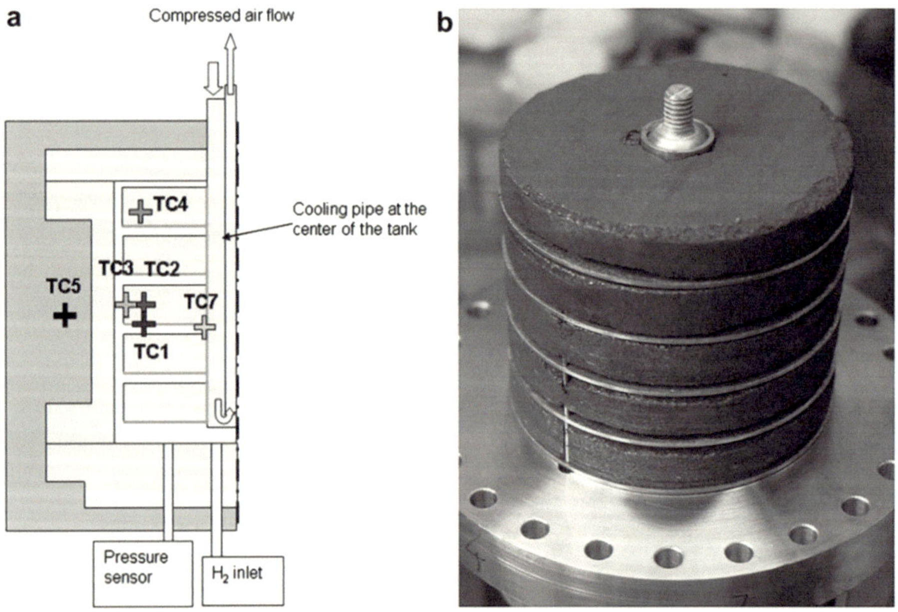

Figure 13.
Sketch of the tank equipped with a copper fin heat exchanger fitting the compacted discs (a) and stack of MgH2 D 5 wt.% ENG (b) [20].

surface area for fuel transport and thus electrochemical reactions, improving the efficiency and performance of the energy storage or conversion devices.

4. Conclusions

Porous media have found wide applications in enhancing heat and mass transfer. The first method to enhance heat transfer by porous media is to increase the thermal conductivity of the material, which is widely used in latent thermal energy storage to improve the heat conductivity of the PCMs. The porous materials used in this regard mainly include two types: graphite-based porous materials (e.g., expanded graphite or graphite foams) and metal foams. The second method for enhancing heat transfer can be realized by enhancing the convective heat transfer, which finds wide applications in heat exchangers, boiling or condensation, and other fields. The enhancement of convective heat transfer is mainly attributed to the larger surface area of porous materials, rich flow channels, formation of local disturbances, and easier formation of bubble nucleation. The enhancement of mass transfer by porous media is, however, attributed to the permeability, which allows effective mass transport inside the porous media. The well-known applications of porous media for mass-transfer enhancement are in crude oil and natural gas extractions and enhanced geothermal energy exploitation. By creating artificial fractures and thus increasing the permeability of the porous media, mass transfer is enhanced in the reservoirs. Another application of porous media for mass-transfer enhancement can be found in hydrogen storage by metal hydride. The porous powder of metals, instead of solid metals, is used to enhance penetration of hydrogen into the interior of materials and undergo chemical reactions with solid materials. Many more applications of porous media in heat and mass transfer are yet to be discovered.

Acknowledgements

The authors gratefully acknowledge the financial support from National Natural Science Foundation of China (Nos. 51606117) and Natural Science Foundation of Shanghai (No. 20ZR1423300).

Conflict of interest

The authors declare no conflict of interest.

Author details

Guojun Yu* and Huijin Xu
Merchant Marine College, Shanghai Maritime University, Shanghai, P.R. China

*Address all correspondence to: gjyu@shmtu.edu.cn

References

[1] Zhong Y, Guo Q, Li S, Shi J, Liu L. Heat transfer enhancement of paraffin wax using graphite foam for thermalenergy storage. Solar Energy Materials and Solar Cells. 2010;**94**:1011-1014. DOI: 10.1016/j.solmat.2010.02.004

[2] Zhong L, Zhang X, Luan Y, Wang G, Feng Y, Feng D. Preparation and thermal properties of porous heterogeneous composite phase change materials based on molten salts/expanded graphite. Solar Energy. 2014;**107**:63-73. DOI: 10.1016/j.solener.2014.05.019

[3] Sari A, Karaipekli A. Thermal conductivity and latent heat thermal energy storage characteristics of paraffin/expanded graphite composite as phase change material. Applied Thermal Engineering. 2007;**27**:1271-1277. DOI: 10.1016/j.applthermaleng.2006.11.004

[4] Xiao X, Zhang P, Li M. Preparation and thermal characterization of paraffin/metal foam composite phase change material. Applied Energy. 2013;**112**:1357-1366. DOI: 10.1016/j.apenergy.2013.04.050

[5] Fleming E, Wen S, Shi L, da Silva AK. Experimental and theoretical analysis of an aluminum foam enhanced phase change thermal storage unit. International Journal of Heat and Mass Transfer. 2015;**82**:273-281. DOI: 10.1016/j.ijheatmasstransfer.2014.11.022

[6] Rehman T-u, Ali HM, Janjua MM, Sajjad U, Yan W-M. A critical review on heat transfer augmentation of phase change materials embedded with porous materials/foams. International Journal of Heat and Mass Transfer. 2019;**135**:649-673. DOI: 10.1016/j.ijheatmasstransfer.2019.02.001

[7] Huang ZF, Nakayama A, Yang K, Yang C, Liu W. Enhancing heat transfer in the core flow by using porous medium insert in a tube. International Journal of Heat and Mass Transfer. 2010;**53**:1164-1174. DOI: 10.1016/j.ijheatmasstransfer.2009.10.038

[8] Ming TZ, Zheng Y, Liu J, Liu C, Liu W, Huang SY. Heat transfer enhancement by filling metal porous medium in central area of tubes. Journal of the Energy Institute. 2010;**83**:17-24. DOI: 10.1179/014426010X12592427711911

[9] Vazifeshenas Y, Sedighi K, Shakeri M. Open cell metal foam as extended coolant surface–fuel cell application. Fuel Cell. 2020;**20**:108-115. DOI: 10.1002/fuce.201800147

[10] Bai P, Tang T, Tang B. Enhanced flow boiling in parallel microchannels with metallic porous coating. Applied Thermal Engineering. 2013;**58**:291-297. DOI: 10.1016/j.applthermaleng.2013.04.067

[11] Ma Y, Huang C, Wang X. Experimental investigation on boiling heat transfer enhanced by gradient aperture porous copper. Applied Thermal Engineering. 2021;**191**:116877. DOI: 10.1016/j.applthermaleng.2021.116877

[12] Bai M. An innovative method for horizontal well completion in tight shale gas reservoirs. In: IADC/SPE Asia Pacific Drilling Technology Conference and Exhibition; 9-11 July 2012; Tianjin, China. SPE; 2012. p. 156509. DOI: 10.2118/156509-MS

[13] Julia FW, Gale SE, Laubach JE, Olson PE, András F. Natural fractures in shale: A review and new observations. AAPG Bulletin. 2014;**98**:2165-2216. DOI: 10.1306/08121413151

[14] US Department of Energy. An Evaluation of Enhanced Geothermal Systems Technology. Washington, DC (United States): EERE Publication and Product Library; 2009. DOI: 10.2172/1217838

[15] Yu G, Li H, Liu C, Cheng W, Xu H. Thermal and hydraulic characteristics of a new proposed flyover-crossing fracture configuration for the enhanced geothermal system. Renewable Energy. 2023;**211**:859-873. DOI: 10.1016/j.renene.2023.04.148

[16] Zhang W, Qu Z, Guo T, Wang Z. Study of the enhanced geothermal system (EGS) heat mining from variably fractured hot dry rock under thermal stress. Renewable Energy. 2019;**143**:855-871. DOI: 10.1016/j.renene.2019.05.054

[17] Bogdanovic B, Reiser A, Schlichte K, Spliethoff B, Tesche B. Thermodynamics and dynamics of the Mg–Fe–H system and its potential for thermochemical thermal energy storage. Journal of Alloys and Compounds. 2002;**345**:77-89. DOI: 10.1016/s0925-8388(02)00308-0

[18] Yu G, Hu Z, Gao Y. Molecular dynamics investigation into the hydrogen adsorption mechanism of magnesium-based material 2Mg-Fe mixture. Energy and Fuels. 2022;**36**:9313-9320. DOI: 10.1021/acs.energyfuels.2c01859

[19] Askri F, Jemni A, Nasrallah SB. Dynamic behavior of metal–hydrogen reactor during hydriding process. International Journal of Hydrogen Energy. 2004;**29**:635-647. DOI: 10.1016/s0360-3199(03)00220-9

[20] Chaise A, de Rango P, Marty P, Fruchart D. Experimental and numerical study of a magnesium hydride tank. International Journal of Hydrogen Energy. 2010;**35**:6311-6322. DOI: 10.1016/j.ijhydene.2010.03.057

Chapter 7

Flow and Heat Transfer in Graded Porous Media and Its Application in Aeroengine Cooling

Xiaohui Bai, Cunliang Liu and Akira Nakayama

Abstract

In recent decades, many studies on flow and heat transfer in porous media have been conducted by researchers to take advantage of its high specific surface area and good heat transfer performance. What is more, the graded porous media have also drawn great attention since the graded arrangement enhances local heat transfer coefficient, so as to improve the overall heat transfer performance and temperature uniformity. The new forms of structure provide a new design of cooling system for some high-power heat sources, such as electronic components, compact heat exchangers, and hot components in aeroengine. In this chapter, the problems of channels filled with vertically graded porous media and axially graded porous media have been introduced, respectively. The flow and heat transfer characteristics of graded porous media have been studied based on porous media theory, considering the parameters of porosity, permeability, pore diameter, etc. The profiles of velocity and temperature vary with different porosity arrangements present. The maximum heat transfer coefficient was obtained for the case of high porosity at the center and low porosity near the wall. Furthermore, the possibility of application in the aeroengine cooling has been discussed.

Keywords: graded porous media, vertically and axially graded, velocity profile, temperature profile, flow and heat transfer characteristics, aeroengine cooling

1. Introduction

Porous media (e.g. metal foam) has drawn great attention due to its potential for application in various scenarios, such as compact heat sinks, regenerative heat exchangers, and combustor-incinerators [1]. Due to its high specific surface area, the metal foam matrix is regarded as a promising material and structure to improve energy absorption and heat transfer in hot component system. In past decades, many relative investigations were conducted theoretically and experimentally and reported for metal foam matrices [2–5].

Functionally graded porous materials are those porous media with a gradual change of porosity (solidity) and pore size. Thanks to the recent advances in manufacturing technologies, the design and fabrication of such non-uniform structures have become possible. Since the concept of functionally graded materials was proposed in the literature [6] in the 1980s, the studies on making continuous changes in the composition, microstructure, porosity, and other properties have been continuously conducted, since these changes improved its mechanical performance or thermal performance in some reasonable ranges. In recent years, some ceramic-made functionally graded materials have been produced by using a position-dependent chemical composition [7].

The graded porous media may achieve a better heat transfer performance since the graded arrangement enhances local heat transfer coefficient, so as to improve the overall heat transfer performance and temperature uniformity. For example, we may consider some functionally graded metal foam structures that are under equal mean porosity. Firstly, the metal foam matrix may be arranged with higher porosity near the wall subject to constant heat flux and lower porosity in the core regions. This arrangement may create a higher velocity close to the wall and lower velocity in the core, which differs from a regular velocity profile in a channel. On the contrary, lower porosity near the wall may result in thicker thermal boundary layer, which prompts the thermal conductivity near the heated wall but weakens convective heat transfer. Therefore, each case of graded porosity arrangement brings favorable and unfavorable factors on heat transfer performance, but an optimal spatial distribution of the porosity may exist.

On the other hand, axially graded porous materials in a channel with lower porosity (i.e. higher metal volume fraction) in the entrance region and higher porosity in the exit region may further enhance the convective heat transfer coefficient in developing region, however, the convective heat transfer in the downstream maybe weaken because the temperature of the coolant air is increased by mixing. On the contrary, the higher porosity (i.e. lower metal volume fraction) in the entrance region and lower porosity in the exit region may enhance the heat transfer performance of the downstream, since the mixing of the coolant air improves its cooling quality in the downstream. Incidentally, the increase of the heat transfer rate downstream accelerated the heat transfer on the wall and improved the temperature uniformity over the wall.

Wang et al. [8] conducted numerical investigation on the heat transfer of a pipe embedded with functionally graded materials with larger diameter particles near the walls. Zheng et al. [9] have numerically evaluated the heat transfer performance of a pipe filled with a porous foam with radially variable sizes of particles under equal porosity. Bai et al. [10] analytically investigated the heat transfer performance of a channel filled with functionally graded foam matrix in a vertical direction from the central line to the heated walls. They concluded that the graded arrangement of foam matrices may achieve a substantial increase in heat transfer coefficient at the expense of an acceptable consume of pressure loss. Subsequently, Iasiello et al. [11] numerically investigated the effects of variable porosity and cell size on the thermal performance of functionally graded foams, whose porosity and cell size vary according to different power-law functions in the direction of normal to the wall. As a result, a 42% increase in Performance Evaluation Criterion (PEC) is found in foams that account for both variable porosity and cell size. Mauro et al. [12] carried out a comprehensive optimization framework to design a graded foam structure by mono-and multi-objective optimization method. They proposed optimum design parameters that achieve the best thermal performance as high as 1.5 performance evaluation criterion (PEC). Recently, Bai et al. [13] considered an occasion of heat transfer in a channel filled with axially graded porous materials to investigate the temperature distribution on the heated walls. They found that the wall temperature variation is quite sensitive to the axial gradient of the local porosity. Furthermore, the functionally graded foams with variable PPI enhance pool boiling heat transfer and that with linearly-variable porosity enhance heat transfer in phase change materials were presented by Xu and Zhao [14] and Yang et al. [15], respectively. In short, the enhancement of heat transfer by exploiting graded porous materials has been comprehensively substantiated.

With the increase of the demand on the large thrust of the engine, the temperature value of gas in front of turbine increases. For the military aircraft, the afterburner is adopted to increase the transient trust of the aeroengine. However, the gas temperature at the afterburning state far exceeds the limitation of the material. In this case, the graded porous structure may be useful to adjust the coolant distribution and to improve the temperature uniformity.

In this chapter, the flow and heat transfer in channels filled with graded porous materials (radially graded and axially graded) have been investigated, and the flow structures and heat transfer characteristics have been revealed for both cases with different parameters. The possibility of the application of graded porous structure in afterburner heat shield has been discussed.

2. Flow and heat transfer in a channel filled with vertically graded porous media

The analytical investigation of heat transfer characteristics of a channel filled with radially graded porous media has been conducted in this section. A series of graded metal foam matrices of the same solidity were arranged with their porosity either increasing or decreasing toward the heated walls following a parabolic function. As a result, the maximum heat transfer coefficient of the channel filled with a functionally graded metal foam matrix was achieved for the case that low porosity near the heated wall and high porosity in the central line, which is 20–50% higher than that of the uniform distribution of metal foam matrix. Therefore, the effectiveness of functionally graded metal foam matrices to achieve higher heat transfer coefficient with an acceptable pressure loss was substantiated.

2.1 Physical model and analysis

As illustrated in **Figure 1**, there are two cases of channels filled with a functionally graded metal matrix, one case has high porosity near the wall and low porosity in the core region of channel, the other case has reversed arrangement. Both upper and lower walls were subject to a constant heat flux q_w. We mainly focus on the region of hydro-dynamically and thermally fully developed flow in these cases, because the developing region, which may take several hydraulic diameters, is short enough due to the addition of the metal foam matrix.

The volume-averaged version of the momentum equation was derived by exploiting the Brinkman-extended-Darcy momentum equation Eq. (1), likewise, the volume-averaged energy equations for the fluid and solid phases were obtained as Eqs. (2) and (3).

$$-\frac{dp}{dx}+\frac{d}{dy}\left(\frac{\mu}{\varepsilon}\frac{du}{dy}\right)-\frac{\mu}{K}u=0 \tag{1}$$

$$\rho_f c_{pf} u\frac{\partial T_f}{\partial x}=\frac{\partial}{\partial y}\left(\varepsilon^* k_f+\varepsilon k_{dis}\right)\frac{\partial T_f}{\partial y}-h_v\left(T_f-T_s\right) \tag{2}$$

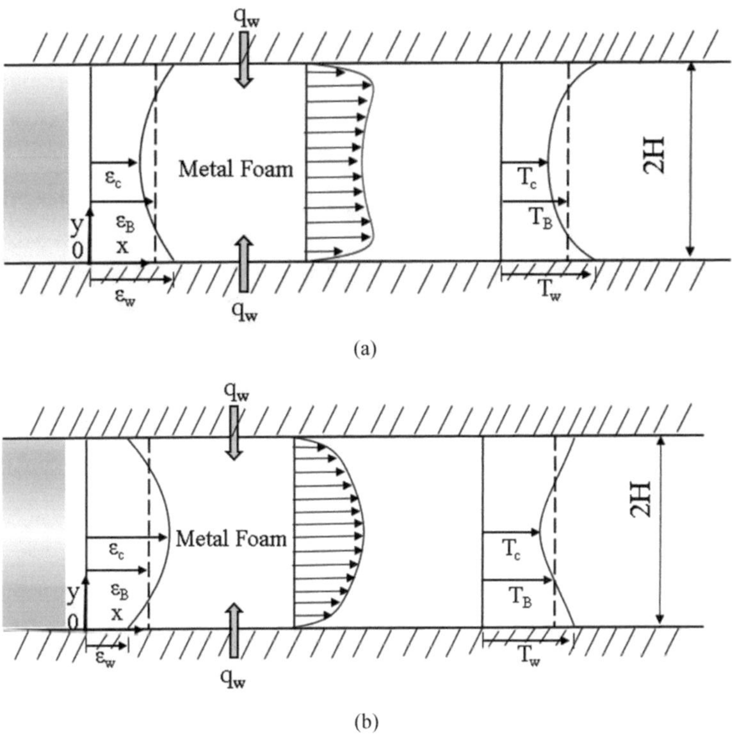

Figure 1.
Physical models of channels with graded porous media. (a) The case with high porosity near the wall and low porosity in the core and (b) the case with low porosity near the wall and high porosity in the core.

$$\frac{\partial}{\partial y}(1-\varepsilon^*)k_s\frac{\partial T_s}{\partial y}+h_v(T_f-T_s)=0 \tag{3}$$

where u is the Dacian bulk velocity, which is related to the local porosity and intrinsic velocity. $\varepsilon(y)$ and $K(\varepsilon)$ are porosity and permeability, respectively, which vary spatially across the channel in vertical direction(y), whose origin is located on the lower heated wall. T_f and T_s are the intrinsic volume average fluid and solid phase temperatures, respectively. h_v is the volumetric interstitial heat transfer coefficient. In general, the thermal dispersion term is often expressed by the gradient diffusion hypothesis with the thermal dispersion conductivity k_{dis} [16]. The concept of the effective porosity ε^* was introduced to account for the tortuosity of the structure, which is theoretically given as

$$\varepsilon^*\equiv\frac{k_s-k_e}{k_s-k_f}\cong\frac{2+\varepsilon}{3} \tag{4}$$

where k_e is the effective stagnant thermal conductivity, which is given as

$$k_e(\varepsilon)=\varepsilon^*k_f+(1-\varepsilon^*)k_s=\frac{2+\varepsilon}{3}k_f+\frac{1-\varepsilon}{3}k_s \tag{5}$$

The two energy equations, namely Eqs. (3) and (4), may be added to yield

$$\rho_f c_{pf} u \frac{\partial T_f}{\partial x} = \frac{\partial}{\partial y}\left(\left(\varepsilon^* k_f + \varepsilon k_{dis}\right)\frac{\partial T_f}{\partial y} + (1-\varepsilon^*)k_s \frac{\partial T_s}{\partial y}\right) \tag{6}$$

Then, Eq. (6) may be reduced under the local thermal equilibrium assumption and neglecting the comparatively small term of the thermal dispersion conductivity k_{dis} for the case of aluminum-air combination [17] to

$$\rho_f c_{pf} u \frac{\partial T}{\partial x} = \frac{\partial}{\partial y}\left(k_e(\varepsilon)\frac{\partial T}{\partial y}\right) \tag{7}$$

Eq. (7) should be solved with the momentum Eq. (1) under the following boundary conditions:

For $y = 0$:

$$u = 0 \text{ and } q_w = -k_e(\varepsilon_w)\frac{\partial T}{\partial y} \tag{8}$$

For $y = H$:

$$\frac{\partial u}{\partial y} = 0 \text{ and } \frac{\partial T}{\partial y} = 0 \tag{9}$$

The porosity of the functionally graded metal foam matrix in a channel varies following a parabolic function across the channel as follows:

$$\varepsilon(y) = \varepsilon_c + (\varepsilon_w - \varepsilon_c)\left(1 - \frac{y}{H}\right)^2 \tag{10}$$

where the subscripts of w and c indicate the wall and the center of the channel, respectively.

As for the permeability, the correlation provided by Calmidi [18] for metal foam was adopted:

$$K(\varepsilon) = 0.000365(1-\varepsilon)^{-0.224}\left(\frac{1.18}{1 - e^{-(1-\varepsilon)/0.04}}\sqrt{\frac{1-\varepsilon}{3\pi}}\right)^{-1.11} H^2 \tag{11}$$

where $K(\varepsilon)/H^2$ is function of the porosity, which typically ranges from 3×10^{-3} to 8×10^{-3}.

Subsequently, the non-dimensional momentum equation and energy equation were solved using a standard Runge-Kutta-Gill scheme, with the following dimensionless boundary conditions:

$$u^*|_{\eta=0} = 0 \tag{12}$$

and

$$\left.\frac{du^*}{d\eta}\right|_{\eta=1} = 0 \tag{13}$$

Once the velocity profile was determined and fed into the energy Eq. (6), the following expression can be yielded by integrating from y = 0 to H

$$H\rho_f c_{p_f} u_B \frac{dT_B}{dx} = q_w \tag{14}$$

where

$$T_B = \frac{\rho_f c_{pf} \int_0^H uTdy}{\rho_f c_{pf} u_B} \tag{15}$$

is the bulk mean temperature. Since $dT_B/dx = \partial T/\partial x$ for the thermally fully developed flow, the energy Eq. (6) simplifies to

$$\frac{u^*}{u*_B} = \frac{d}{d\eta}\left(\frac{k_e(\varepsilon)}{k_e(\varepsilon_B)}\frac{dT^*}{d\eta}\right) \tag{16}$$

where

$$u*_B = \int_0^1 u^* d\eta \tag{17}$$

and

$$T^* = \frac{k_e(\varepsilon_B)(T - T_w)}{q_w H} \tag{18}$$

are the dimensionless bulk mean velocity and local dimensionless temperature, respectively. Integrate the Eq. (16) from $\eta = 0$ can attain

$$\frac{dT^*}{d\eta} = \frac{\int_0^\eta \frac{u^*}{u*_B} d\eta - 1}{\frac{k_e(\varepsilon)}{k_e(\varepsilon_B)}} \tag{19}$$

The dimensionless temperature can be found by further integrating the foregoing Eq. (19) by using the Runge-Kutta-Gill scheme. The following boundary conditions on the temperature and its gradient on the wall are implemented:

$$T^*|_{\eta=0} = 0 \text{ and } \left.\frac{dT^*}{d\eta}\right|_{\eta=0} = -\frac{k_e(\varepsilon_B)}{k_e(\varepsilon_w)} \tag{20}$$

Thus, the Nusselt number of our primary interest can be obtained from

$$Nu_H \equiv \frac{q_w H}{k_e(\varepsilon_B)(T_w - T_B)} = -\frac{1}{\int_0^1 \frac{u^*}{u*_B} T^* d\eta} = \frac{1}{\int_0^1 \frac{\left(\int_0^\eta \frac{u^*}{u*_B} d\eta\right)\left(\int_0^\eta \frac{u^*}{u*_B} d\eta - 1\right)}{\frac{k_e(\varepsilon)}{k_e(\varepsilon_B)}} d\eta - T^*|_{\eta=1}} \tag{21}$$

The definition of Nu_H contains $k_e(\varepsilon_B)$ accounted for the effect of increase in the effective thermal conductivity on the Nusselt number. It is meaningful to compare

Nu_H for all cases of different weights, even for the case of the channel without a metal foam, based on the definition given by Eq. (21).

2.2 Velocity and temperature profiles

The dimensionless velocity profiles obtained for the metal foams of the same weight of ε_B = 0.933 were presented in **Figure 2**. The wall porosity ε_w varies from 0.840–0.980, and the case of ε_w = 0.933 corresponds to the case of uniform porosity metal foam. As we can see, the increase in the ε_w creates a higher velocity value close to the wall and lower velocity with a concave profile in the core region.

To the contrary, the decrease of the ε_w results in a higher velocity value in the core region and lower velocity value near the wall comparing with the velocity distribution for the case of the uniform porosity. Naturally, more metal near the wall has favorable influence on conductive wall heat transfer, but more metal may block the near-wall flow, decreasing the flow velocity. Therefore, these favorable and unfavorable factors form a trade-off for the heat transfer enhancement.

The corresponding dimensionless temperature profile for the case of $k_s/k_f = 10^4$ was shown in **Figure 3**, which assumes a combination of air and aluminum.

As can be seen, the dimensionless temperature profile of the uniform porosity case of $\varepsilon_w = \varepsilon_w$ = 0.933 has been presented and indicated. As increasing the value of ε_w, the dimensionless temperature gradient at the wall become steeper, which can also be expected from Eq. (20). While decreasing the ε_w, the dimensionless temperature profile near the wall changes its curvature to negative one, while the protrusion of the dimensionless temperature in the core becomes more extensive.

2.3 Heat transfer characteristics

Figure 4 presents the curves of Nu_H against the porosity distribution for the given bulk mean porosity ε_B = 0.900, 0.933, and 0.950. The level of Nu_H for the channel

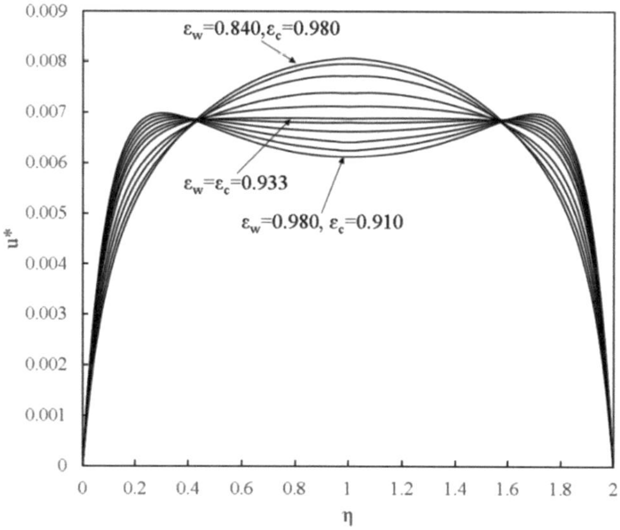

Figure 2.
Dimensionless velocity profiles.

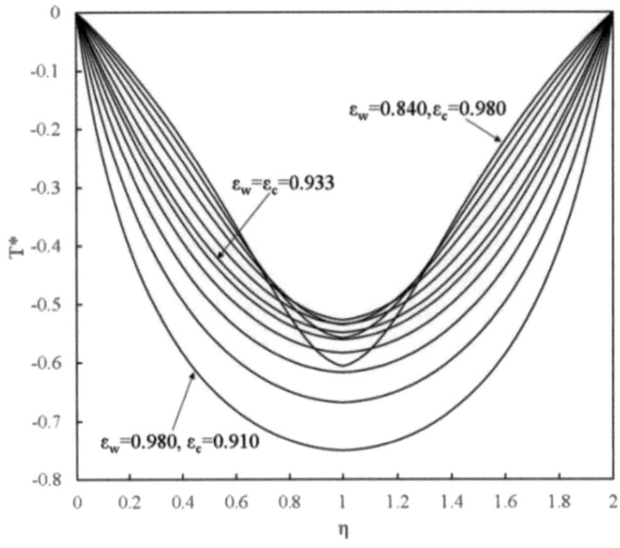

Figure 3.
Dimensionless temperature profiles.

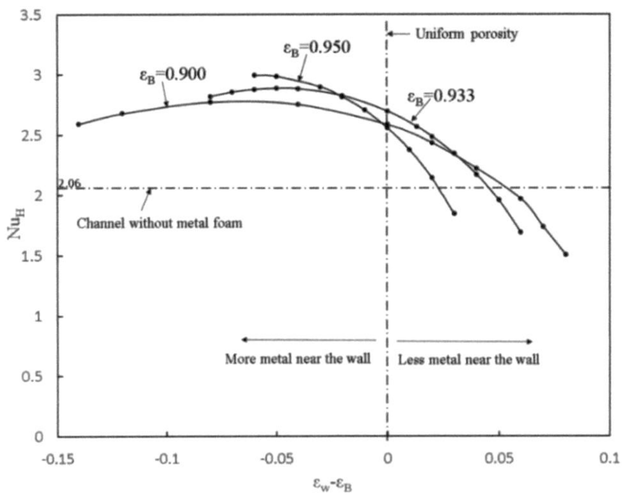

Figure 4.
Nusselt number Nu_H against $(\varepsilon_w - \varepsilon_B)$.

without a metal foam, namely, Nu_H = 2.06, was also indicated. The figure indicates that the increase in the effective thermal conductivity enables the convective heat transfer coefficient in a channel filled with a metal foam to increase about 20–50% more than 2.06 $k_e(\varepsilon_B)/H$. However, for the cases of positive ε_w–ε_B, the values of Nu_H go below the value 2.06.

As a result, the maximum values of Nu_H are found in the negative range of ε_w–ε_B, where the porosity of the functionally graded metal foam decreases toward the wall (i.e. more metal near the wall).

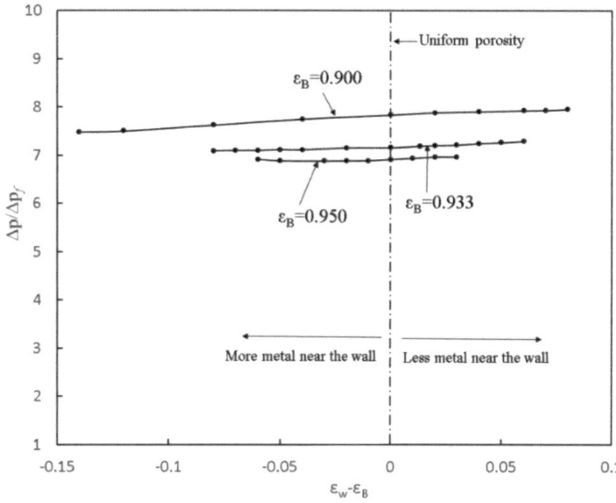

Figure 5.
Pressure drop ratio $\Delta p/\Delta p_f$ against $(\varepsilon_w - \varepsilon_B)$.

In comparison with the case of the channel without a metal foam, the addition of the metal foam achieves high heat transfer coefficient but at expense of a larger pressure drop, or pressure gradient $(-dp/dx)$, which can be related to its pumping power per unit axial length $Hu_B(-dp/dx)$ as

$$-\frac{dp}{dx} = \sqrt{\frac{1}{u*_B}\frac{\mu}{H^3}\left(Hu_B\left(-\frac{dp}{dx}\right)\right)} \tag{22}$$

Thus, the pressure drop ratio in the channel with a metal foam to that of an empty channel (where $u^*_f = 1/3$) under equal pumping power is given by

$$\Delta p/\Delta p_f = 1/\sqrt{3u*_B} \tag{23}$$

As shown in **Figure 5**, the pressure drop ratios for the corresponding three cases are presented. In comparison to the case of the channel without a metal foam, the addition of the metal foam matrix results in about 7–8 times increase in pressure drop under equal pumping power. Moreover, the bulk mean porosity of ε_B brings more sensitive effects than $(\varepsilon_w - \varepsilon_B)$.

2.4 Conclusions

In this section, the heat transfer characteristics of forced convection in channels filled with functionally graded metal foam in vertical direction normal to the heated walls have been analytically investigated. The fully developed set of Brinkman-extended Darcy momentum equation and the energy equation associated with local thermal equilibrium assumption were taken into account. The porosity of parabolic distribution was assumed across the channel height. The integrated energy equation associated with momentum equation was solved by using the Rung-Kutta-Gill integration scheme. According to the analytical results, for the metal foams of the same weight, the maximum heat transfer coefficient can be obtained for the case of the high porosity in the core and low porosity near the wall, at the expense of an acceptable increase in pressure drop.

3. Flow and heat transfer in a channel filled with axiallygraded porous media

In this section, the problem of flow and heat transfer in a channel filled with axially graded porous materials based on local thermal non-equilibrium has been investigated to find a way to control its wall temperature distribution [13]. A series of axially graded foams under an equal solid fraction were considered. The porosity decreasing along the downstream of the channel following a power function of axial coordinate with an exponent of n was assumed. The Brinkman-Forchheimer extended Darcy model and the local thermal non-equilibrium assumption were exploited to obtain the velocity profiles and temperature distributions on the heated walls, respectively.

3.1 Physical model and analysis

A channel of height 2H filled with a fluid-saturated axially graded porous media subject to constant heat flux is presented in **Figure 6**.

The indexes of δ_f and δ_s represent thermal boundary layer thicknesses for fluid phase and solid phase, respectively. Both the fluid phase thermal boundary layer and solid phase thermal boundary layer grow from the entrance to the downstream. However, the thermal boundary layer of solid grows thicker than that of the fluid due to the higher thermal conductivity of the solid than that of the fluid. In the present study, the thermally developing region of the entrance is focused so that δ_f and δ_s are less than half the channel height, H.

In the present case, the Brinkman-Forchheimer-extended Darcy model is still valid since the velocity change in such an axially variable porosity-arranged porous media is so moderate that the convective inertia terms are almost negligible. Hence, the velocity field can be determined from the momentum and continuity equations as follows:

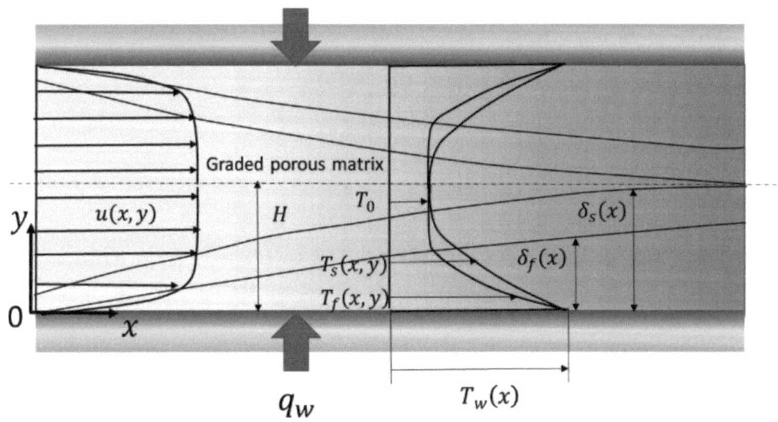

Figure 6.
Channel model filled with an axially graded porous medium.

$$-\frac{dp}{dx}+\frac{\mu_f}{\varepsilon(x)}\frac{\partial^2 u}{\partial y^2}-\frac{\mu_f}{K(x)}u-\rho_f b(x)u^2=0 \tag{24}$$

$$\frac{\partial u}{\partial x}+\frac{\partial v}{\partial y}=0 \tag{25}$$

where u and v are the local volume average velocity components. p is the intrinsic pressure, μ_f is the fluid viscosity, $K(x)$ is the permeability, and b is the inertial coefficient. Moreover, x is the axial coordinate while y is the coordinate measured vertically from the lower wall surface of the channel.

Following Amiri and Vafai [19], the energy equations for the fluid phase and the solid phase can be given as

$$\rho_f c_{p_f}\frac{\partial uT_f}{\partial x}+\rho_f c_{p_f}\frac{\partial vT_f}{\partial y}=\frac{\partial}{\partial y}k_{fe}(x)\frac{\partial T_f}{\partial y}-h_v(x)\left(T_f-T_s\right) \tag{26}$$

$$\frac{\partial}{\partial y}k_{se}(x)\frac{\partial T_s}{\partial y}+h_v(x)\left(T_f-T_s\right)=0 \tag{27}$$

where k_{fe} and k_{se} are the effective thermal conductivities of the fluid and the solid, respectively. h_v denotes the interstitial volumetric heat transfer coefficient.

Considering the symmetry of the geometry, the lower half of the channel is analyzed with corresponding boundary conditions given as follows:

$$x=0: T_f-T_0=\text{const.} \tag{28}$$

$$y=0: u=v=0, T_f=T_s=T_w(x) \text{ and } q_w=-k_{fe}\left.\frac{\partial T_f}{\partial y}\right|_{y=0}-k_{se}\left.\frac{\partial T_s}{\partial y}\right|_{y=0}=\text{const.} \tag{29}$$

$$y=H: v=0, \left.\frac{\partial T_f}{\partial y}\right|_{y=H}=\left.\frac{\partial T_s}{\partial y}\right|_{y=H}=0 \tag{30}$$

where q_w is the wall heat flux. $T_w(x)$ is the wall temperature.

According to Yi et al. [20], the Eq. (24) can be normalized by using the constant bulk velocity u_B with axially variable porosity and permeability as

$$\frac{\partial^2}{\partial \eta^2}\left(\frac{u}{u_B}\right)-\left(\frac{\varepsilon(x)}{Da(x)}\right)\left(\frac{u}{u_B}\right)-Re_b(x)\left(\frac{u}{u_B}\right)^2=-\left(-\frac{\varepsilon H^2}{\mu_f u_B}\frac{dp}{dx}\right) \tag{31}$$

where

$$\eta=\frac{y}{H} \tag{32}$$

$$Da(x)=\frac{K(x)}{H^2}\quad \text{(Darcy number)} \tag{33}$$

$$Re_b(x)=\frac{\varepsilon(x)\rho_f b(x)H^2 u_B}{\mu_f} \tag{34}$$

where $Re_b(x)$ is the Forchheimer Reynolds number based on the scale of length bH^2. The following velocity profile $u(x,\eta)$ may be assumed with an unknown parameter $\zeta(x)$:

$$\frac{u}{u_B} = \frac{\zeta+1}{\zeta}\left(1-(1-\eta)^{\zeta}\right) \tag{35}$$

which automatically satisfies the boundary conditions (29) and $\int_0^1\left(\frac{u}{u_B}\right)d\eta = 1$. It is obvious that the case of $\zeta = 2$ in Eq. (35) denotes the fully developed profile in an empty channel flow, while for the case of $\zeta \to \infty$, the velocity profile becomes the Darcy(plug) flow.

Evaluating Eq. (31) on the wall (at $\eta = 0$) using the foregoing velocity profile (35), one obtains

$$-\frac{\varepsilon H^2}{\mu_f u_B}\frac{dp}{dx} = \zeta^2 - 1 \tag{36}$$

Thus, Eq. (31) may be integrated across the channel from $\eta = 0$ to 1 as

$$-\frac{\partial}{\partial\eta}\left(\frac{u}{u_B}\right)\Big|_{\eta=0} - \left(\frac{\varepsilon}{Da}\right)\int_0^1\left(\frac{u}{u_B}\right)d\eta - Re_b\int_0^1\left(\frac{u}{u_B}\right)^2 d\eta = -(\zeta^2-1) \tag{37}$$

Based on the Eq. (37) associated with the Eq. (31), a cubic equation for the unknown parameter ζ can be yielded as follows:

$$2\zeta^3 - \zeta^2 - \left(2\frac{\varepsilon}{Da} + 2Re_b + 5\right)\zeta - \left(\frac{\varepsilon}{Da} + 2Re_b + 2\right) = 0 \tag{38}$$

The local value of the $\zeta(\varepsilon/Da,\ Re_b)$ in this cubic equation can be easily evaluated by substituting the local values of $\varepsilon(x)/Da(x)$ and $Re_b(x)$. As for the root of the foregoing cubic equation, a highly accurate expression in a closed form based on the two asymptotes was obtained by Yi et al. [20]:

$$\zeta\left(\left(\frac{\varepsilon}{Da}\right),\ Re_b\right) = \frac{Re_b\,\frac{3+\sqrt{25+16\left(Re_b+\frac{\varepsilon}{Da}\right)}}{4} + \left(\frac{\varepsilon}{Da}\right)\frac{1+\sqrt{9+4\left(\frac{\varepsilon}{Da}+Re_b\right)}}{2}}{\left(\frac{\varepsilon}{Da}\right) + Re_b} \tag{39}$$

Eq. (39) reduces to the two distinct exact asymptotes as:

$$\zeta\left(\frac{\varepsilon}{Da},0\right) = \frac{1+\sqrt{9+4\frac{\varepsilon}{Da}}}{2} \tag{40}$$

and

$$\zeta(0,\ Re_b) = \frac{3+\sqrt{25+16\,Re_b}}{4} \tag{41}$$

Naturally, the velocity profile function with Eqs. (40) and (41) gives $\frac{u}{u_B} = \frac{3}{2}\left(1-(1-\eta)^2\right)$ for $\frac{Da}{\varepsilon} \to \infty$ (or $Re_b \to 0$) (i.e. empty channel) and $\frac{u}{u_B} = 1$ for $\frac{Da}{\varepsilon} \to 0$ (or $Re_b \to \infty$) (i.e. Darcy flow).

Once $\zeta(x)$ is determined, the local friction coefficient may be estimated from

$$\frac{H}{\rho_f u_B{}^2}\left(-\frac{dp}{dx}\right) = \frac{\zeta^2 - 1}{\varepsilon Re_H} \tag{42}$$

where

$$Re_H = \frac{\rho_f u_B H}{\mu_f} \tag{43}$$

is the Reynolds number based on u_B and H.

The vertical velocity component $v(x, \eta)$ may be determined by substituting Eq. (31) into the continuity Eq. (25) as

$$\begin{aligned} \frac{v}{u_B} &= -H\frac{\partial}{\partial x}\int_0^\eta \frac{u}{u_B}d\eta = -H\frac{\partial}{\partial x}\frac{\zeta+1}{\zeta}\int_0^\eta \left(1-(1-\eta)^\zeta\right)d\eta \\ &= -\frac{H}{\zeta^2}\frac{d\zeta}{dx}(1-\eta)\left(1-(1-\eta)^\zeta(1-\zeta\ ln\,(1-\eta))\right) \end{aligned} \tag{44}$$

which satisfies the boundary conditions (29). It is interesting to note that the velocity component v in the lower half of the channel is negative for the channel of $\frac{d\zeta}{dx} > 0$, which is met when either $\frac{\varepsilon}{Da} = \frac{\varepsilon(x)H^2}{K(x)}$ or $Re_b = \frac{\varepsilon(x)\rho_f b(x) H^2 u_B}{\mu_f}$ increases downstream, while the vertical velocity component v vanishes in the case of channels filled with uniform porous materials (i.e. $\frac{d\zeta}{dx} = 0$).

For the thermally fully developed flows in a channel filled with a uniform porous medium, Nakayama et al. [4] obtained a temperature field under the L.T.E. assumption as

$$\begin{aligned} \frac{T_f - T_w}{T_{fB} - T_w} &= \frac{T_s - T_w}{T_{fB} - T_w} \\ &= \frac{3(\zeta+3)(2\zeta+3)}{2\zeta(\zeta+2)(2\zeta^2+13\zeta+17)}\left((\zeta+1)(\zeta+2)\left(1-(1-\eta)^2\right)\right. \\ &\quad \left. -2\left(1-(1-\eta)^{\zeta+2}\right)\right) \end{aligned} \tag{45}$$

For the case of the channel filled with axially graded porous materials, analytical expressions for its wall temperature under the L.T.N.E. assumption were obtained by exploiting the above temperature profile (Eq. (45)) and velocity profile (Eq. (35)), which prevail across both fluid and solid thermal boundary layers ($\delta_f < \delta_s < H$):

$$\frac{T_f - T_0}{T_w - T_0} = \begin{cases} 1 - \dfrac{(\zeta+1)(\zeta+2)\left(1-\left(1-\frac{\eta}{\delta_f{}^*}\right)^2\right) - 2\left(1-\left(1-\frac{\eta}{\delta_f{}^*}\right)^{\zeta+2}\right)}{\zeta(\zeta+3)} & : 0 \le \eta \le \delta_f{}^* \\ 0 : \delta_f{}^* \le \eta \le 1 \end{cases} \tag{46}$$

and

$$\frac{T_s - T_0}{T_w - T_0} = \begin{cases} 1 - \dfrac{(\zeta+1)(\zeta+2)\left(1-\left(1-\frac{\eta}{\delta_s{}^*}\right)^2\right) - 2\left(1-\left(1-\frac{\eta}{\delta_s{}^*}\right)^{\zeta+2}\right)}{\zeta(\zeta+3)} & : 0 \le \eta \le \delta_s{}^* \\ 0 : \delta_s{}^* \le \eta \le 1 \end{cases} \tag{47}$$

where

$$\delta_f{}^* = \delta_f / H \tag{48}$$

and

$$\delta_s{}^* = \delta_s / H \tag{49}$$

These profiles satisfied the boundary conditions (29) and (30) and naturally recovered the fully-developed profile given by Eq. (45) as both $\delta_f{}^*$ and $\delta_s{}^*$ grow downstream to reach unity. The fluid bulk mean temperature T_{fB} in the entrance region may be estimated using the profiles (35) and (46) as

$$\begin{aligned} \frac{T_{fB} - T_0}{T_w - T_0} &\equiv \int_0^1 \frac{u}{u_B}\left(\frac{T_f - T_0}{T_w - T_0}\right) dy^* \\ &= \int_0^{\delta_f{}^*} \frac{\zeta+1}{\zeta}\left(1-(1-y^*)^{\zeta}\right)\left(1 - \frac{(\zeta+1)(\zeta+2)\left(1-\left(1-\frac{y^*}{\delta_f{}^*}\right)^2\right) - 2\left(1-\left(1-\frac{y^*}{\delta_f{}^*}\right)^{\zeta+2}\right)}{\zeta(\zeta+3)}\right) dy^* \\ &\cong \int_0^{\min\left(\frac{1}{\zeta},\delta_f{}^*\right)} \frac{\zeta+1}{\zeta}(\zeta y^*)\left(1 - \frac{(\zeta+1)(\zeta+2)\left(1-\left(1-\frac{y^*}{\delta_f{}^*}\right)^2\right) - 2\left(1-\left(1-\frac{y^*}{\delta_f{}^*}\right)^{\zeta+2}\right)}{\zeta(\zeta+3)}\right) dy^* \\ &\quad + \int_{\min\left(\frac{1}{\zeta},\delta_f{}^*\right)}^{\delta_f{}^*} \frac{\zeta+1}{\zeta}\left(1 - \frac{(\zeta+1)(\zeta+2)\left(1-\left(1-\frac{y^*}{\delta_f{}^*}\right)^2\right) - 2\left(1-\left(1-\frac{y^*}{\delta_f{}^*}\right)^{\zeta+2}\right)}{\zeta(\zeta+3)}\right) dy^* \\ &= \frac{(\zeta+1)(\zeta+5)(\zeta^2+5\zeta+10)}{12(\zeta+3)^2(\zeta+4)}\delta_f{}^{*2} \\ &\quad - \frac{(\zeta+1)^2(\zeta+2)}{12\zeta(\zeta+3)}\left(1 - \frac{\min\left(\frac{1}{\zeta},\delta_f{}^*\right)}{\delta_f{}^*}\right)^4\left(1 - \frac{24\left(1 - \frac{\min\left(\frac{1}{\zeta},\delta_f{}^*\right)}{\delta_f{}^*}\right)^{\zeta}}{(\zeta+1)(\zeta+2)(\zeta+3)(\zeta+4)}\right)\delta_f{}^{*2} \end{aligned} \tag{50}$$

Since the fluid phase thermal boundary layer is thinner in this entrance developing region, the velocity field across the thermal boundary layer may be described by a linearized velocity profile.

The interstitial heat transfer term $h_v(x)(T_f - T_s)$ can be canceled by adding the energy Eqs. (26) and (27). Moreover, integrating the resulting equation across the channel and over the axial distance from 0 to x with the boundary conditions (28) and (29), the energy balance relationship can be obtained as follows:

$$\rho_f c_{p_f} u_B H (T_{fB} - T_0) = q_w x \tag{51}$$

where

$$q_w = -k_{fe} \frac{\partial T_f}{\partial y}\bigg|_{y=0} - k_{se} \frac{\partial T_s}{\partial y}\bigg|_{y=0} = \frac{(T_w - T_0)}{H} \frac{2(\zeta+2)}{(\zeta+3)} \left(\frac{k_{fe}}{\delta_f{}^*} + \frac{k_{se}}{\delta_s{}^*} \right) \tag{52}$$

Furthermore, integrating the solid phase heat conduction Eq. (4) from y = 0 to H with the boundary condition (30) may yield

$$-k_{se}(x) \frac{\partial T_s}{\partial y}\rfloor_{y=0} + h_v(x) \int_0^H (T_f - T_s) dy = 0 \tag{53}$$

The temperature profile functions (46) and (47) are fed into the foregoing equation to find

$$\delta_s{}^* \left(\delta_s{}^* - \delta_f{}^* \right) = \frac{6}{Bi(x)} \frac{(\zeta+2)(\zeta+3)}{(\zeta^2 + 6\zeta + 11)} \tag{54}$$

which gives the solid phase thermal boundary layer thickness $\delta_s{}^*$:

$$\frac{\delta_s{}^*}{\delta_f{}^*} = \frac{1}{2} \left(1 + \sqrt{1 + \frac{24}{Bi(x)} \frac{(\zeta+2)(\zeta+3)}{(\zeta^2 + 6\zeta + 11)\delta_f{}^{*2}}} \right) \tag{55}$$

where

$$Bi(x) = \frac{h_v(x) H^2}{k_{se}(x)} \tag{56}$$

is the local Biot number accounts for the interstitial convective conductance and the solid phase conductance.

Eqs. (50)–(52), and (55) are combined to eliminate $(T_{fB} - T_0)$, q_w and $\delta_s{}^*$ resulting in the following relationship between $\delta_f{}^*$ and ξ:

$$\xi = \frac{\frac{(\zeta+1)(\zeta+5)\left(\zeta^2+5\zeta+10\right)}{12(\zeta+3)^2(\zeta+4)} \delta_f{}^{*2} - \frac{(\zeta+1)^2(\zeta+2)}{12\zeta(\zeta+3)} \left(1 - \frac{min\left(\frac{1}{\zeta}, \delta_f{}^*\right)}{\delta_f{}^*} \right)^4 \left(1 - \frac{24\left(1 - \frac{min\left(\frac{1}{\zeta}, \delta_f{}^*\right)}{\delta_f{}^*}\right)^\zeta}{(\zeta+1)(\zeta+2)(\zeta+3)(\zeta+4)} \right) \delta_f{}^{*2}}{\frac{2(\zeta+2)}{(\zeta+3)\delta_f{}^*} \left(\frac{k_{fe}(x)}{k_{ref}} + \frac{k_{se}(x)}{k_{ref}} \frac{2}{1 + \sqrt{1 + \frac{24}{Bi} \frac{(\zeta+2)(\zeta+3)}{(\zeta^2+6\zeta+11)\delta_f{}^{*2}}}} \right)} \tag{57}$$

where

$$\xi = \frac{(x/H)}{\left(\frac{\rho_f c_{p_f} u_B H}{k_{ref}}\right)} \tag{58}$$

is the Graetz number. Moreover, k_{ref} is a reference thermal conductivity.

For given properties of the graded porous materials, namely $\frac{\varepsilon(x)}{Da(x)}$ and $Re_b(x) = \frac{\varepsilon(x)\rho_f b(x) H^2 u_B}{\mu_f}$, the velocity profile parameter $\zeta(x)$ can be determined from Eq. (37). Once $\zeta(x)$ is known, the fluid phase boundary layer thickness $\delta_f{}^*$ can be found by using Eq. (57) as a function of $\xi = \frac{(x/H)}{\left(\frac{\rho_f c_{p_f} u_B H}{k_{ref}}\right)}$ for given local values of $\frac{k_{fe}(x)}{k_{ref}}, \frac{k_{se}(x)}{k_{ref}}, Bi(x) = \frac{h_v(x)H^2}{k_{se}(x)}$. With $\zeta(x)$ and $\delta_f{}^*(x)$ thus determined, the development of the dimensionless wall temperature $\frac{k_{ref}(T_w(x)-T_0)}{Hq_w}$ of great interest may readily be found from

$$\frac{k_{ref}(T_w(x)-T_0)}{Hq_w} - \frac{\frac{(\zeta+3)}{2(\zeta+2)}\delta_f{}^*}{\frac{k_{fe}(x)}{k_{ref}} + \frac{k_{se}(x)}{k_{ref}}\frac{2}{1+\sqrt{1+\frac{24}{Bi(x)}\frac{(\zeta+2)(\zeta+3)}{(\zeta^2+6\zeta+11)\delta_f{}^{*2}}}}} \tag{59}$$

which is based on Eqs. (52) and (55). Correspondingly, the local Nusselt number may be calculated from

$$\begin{aligned} Nu_H &\equiv \frac{q_w H}{(T_w - T_{fB})k_{ref}} = 1/\left(\frac{k_{ref}(T_w(x)-T_0)}{Hq_w} - \frac{k_{ref}(T_{fB}(x)-T_0)}{Hq_w}\right) \\ &= \frac{1}{\frac{\frac{(\zeta+3)}{2(\zeta+2)}\delta_f{}^*}{\frac{k_{fe}(x)}{k_{ref}} + \frac{k_{se}(x)}{k_{ref}}\frac{2}{1+\sqrt{1+\frac{24}{Bi(x)}\frac{(\zeta+2)(\zeta+3)}{(\zeta^2+6\zeta+11)\delta_f{}^{*2}}}}} - \xi} \end{aligned} \tag{60}$$

3.2 Porosity variation and velocity profiles

As described in Eq. (61), the porosity of the graded metal foam in the channel of length L was assumed to vary axially according to the function as follows:

$$\varepsilon(x) = 1 - (1-\varepsilon_{ref})(1+n)\left(\frac{x}{L}\right)^n \tag{61}$$

such that the solidity remains constant irrespective of the value of the exponent n:

$$\frac{2H\int_0^L (1-\varepsilon(x))dx}{(2HL)} = (1-\varepsilon_{ref}) : \text{Solidity} \tag{62}$$

where ε_{ref} is the mean porosity of the channel. The porosity decreases (i.e. the solidity increases) steeply downstream as increasing the value of n from 0.

The correlations for the permeability K and interstitial heat transfer coefficients h_v can be found in Calmid [18], in which the hydrodynamic and thermal characteristics of flows in metal foams were investigated. A typical correlation for the permeability of metal foam was proposed in their publications:

$$Da\left(\varepsilon;\frac{d_m}{H}\right)=\frac{K}{H^2}=0.00073(1-\varepsilon)^{-0.224}\left(\frac{1.18}{1-e^{-(1-\varepsilon)/0.04}}\sqrt{\frac{1-\varepsilon}{3\pi}}\right)^{-1.11}\left(\frac{d_m}{H}\right)^2 \tag{63}$$

In order to overcome the deficiency around the inlet in the Calmid correlation, a simple power law relationship is proposed, which is as follows:

$$Da=Da_{ref}\left(\frac{\varepsilon}{\varepsilon_{ref}}\right)^m \tag{64}$$

Wong et al. [21] theoretically recommended the range of the exponent m in three-dimensional porous media. Furthermore, the power law (64) with m = 4 and $Da_{ref}=Da\left(\varepsilon_{ref};\frac{d_m}{H}\right)$ was expressed as

$$Da=\left(0.00073\left(1-\varepsilon_{ref}\right)^{-0.224}\left(\frac{1.18}{1-e^{-\left(1-\varepsilon_{ref}\right)/0.04}}\sqrt{\frac{1-\varepsilon}{3\pi}}\right)^{-1.11}\left(\frac{d_m}{H}\right)^2\right)\left(\frac{\varepsilon}{\varepsilon_{ref}}\right)^4 \tag{65}$$

gives the maximum value at the inlet and then roughly follows the urves based on the Calmid correlation, which continuously decreases downstream.

Figure 7a and **b** illustrate the axial development of the velocity field whose porosity varies following Eq. (61) with Da (x) evaluated from the power law (65) for the case of $\varepsilon_{ref}=0.8$, $d_m/H=0.1, L/H=10,\ Re_H=1500,$ and n = 1.5.

The streamwise velocity profiles u/u_B clearly indicate that the viscous boundary layer becomes "thinner" downstream, as both porosity and permeability decrease downstream. The corresponding normal velocity component $-v/u_B$ at the same locations are presented in **Figure 7b**. **Figure 7a** and **b** together suggest that some part of the fluid mass is shifted from the core toward the heated wall. This weak secondary flow results in a comparatively higher streamwise velocity layer near the wall, which is favorable in view of cooling the heated wall.

3.3 Boundary layer thickness and wall temperature variation

It is found that the wall temperature and Nusselt number variations are quite insensitive to the difference in the local Darcy number observed in the above two correlations. This may be attributed to the forgetfulness of the inlet conditions, which is inherent to the boundary layer theory. Therefore, only the results based on the Calmid correlation (59) are presented.

A series of calculations were conducted analytically and numerically for forced convective flows in a channel filled with a graded metal foam under equal solidity, for

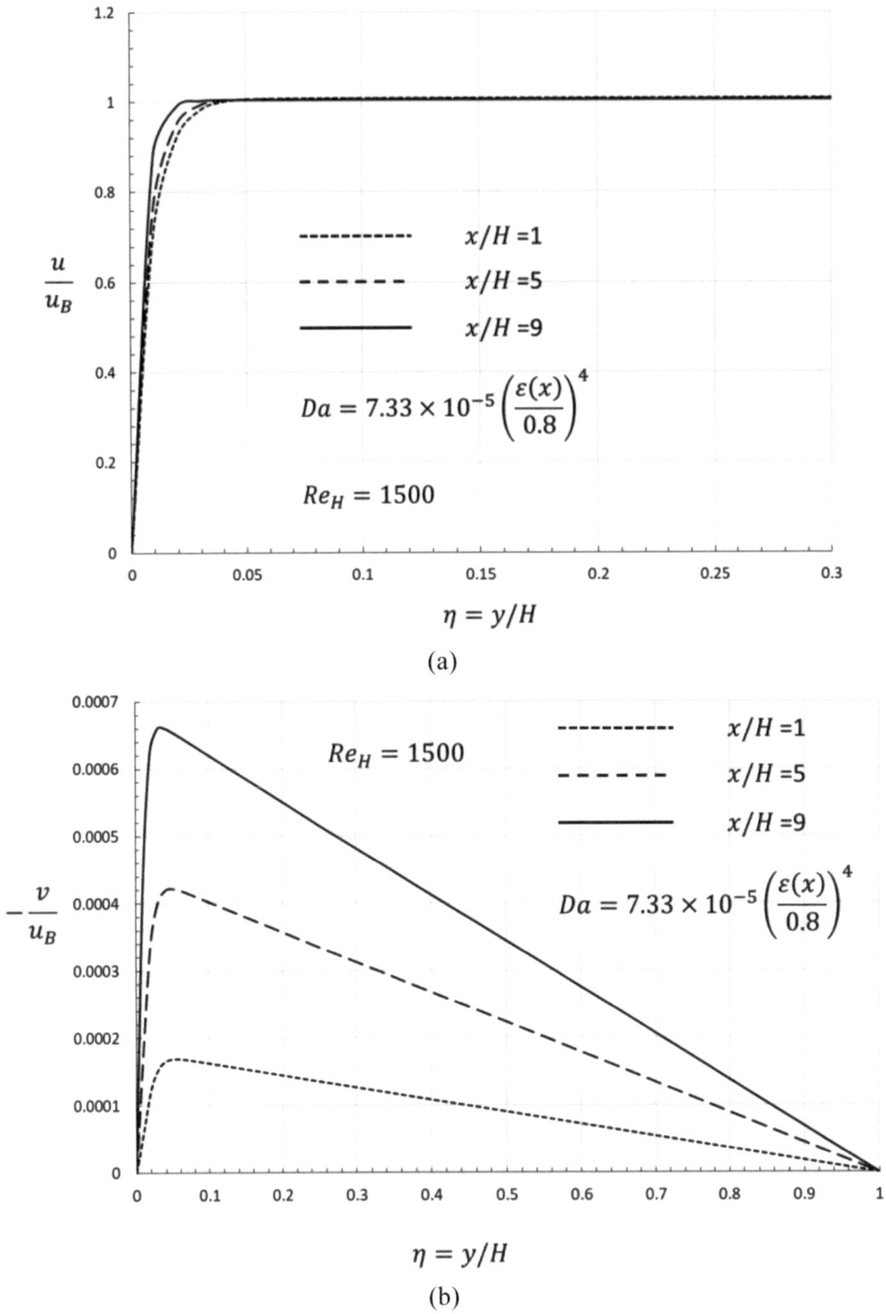

Figure 7.
Axial development of velocity field. (a) Streamwise velocity component and (b) normal velocity components.

the cases of $\varepsilon_{ref} = 0.8, d_m/H = 0.1, L/H = 10$, $Re_H = 1500, Pr = 0.7$, and $\frac{k_f}{k_s} = 0.01$. The exponent n for controlling the gradient of the porosity was set to 0, 0.5, 1.0, and 1.5, respectively. For given exponent n and other parameters, the velocity shape parameter $\zeta(x)$ is evaluated using Eq. (39) to determine the velocity field. The local value of $\delta_f{}^*(x) = \delta_f/H$ also can be found by using a standard shooting method once the $\zeta(x)$ is locally determined, and then some other quantities of interest, such as the wall temperature, can readily be provided.

As illustrated in **Figure 8**, the thermal boundary layer thickness of the solid is always greater than that of the fluid phase due to the thermal conductivity of the solid phase is much greater. In fact, the thermal boundary layer of the fluid phase just started to grow at the inlet, namely $\delta_f\big|_{x=0} = 0$, while the boundary layer thickness of solid phase is non-zero as given by $\delta_s = \sqrt{\frac{6}{Bi}\frac{(\zeta+2)(\zeta+3)}{(\zeta^2+6\zeta+11)}}H$. Therefore, the boundary layer thickness ratio δ_f/δ_s increases from zero at the inlet along downstream. Since $\delta_f/\delta_s = 1$ is required for both fluid and solid temperature profiles to be identical, the L. T.E. assumption fails over a substantial distance from the inlet as can be seen from **Figure 8**.

The effects of the exponent n on the wall temperature development are illustrated in **Figure 9**.

A typical thermal boundary layer development can be observed in **Figure 9** for the case of uniform porosity n = 0, that is, under the condition of constant heat flux on the wall, the wall temperature increases from the inlet in proportion to $x^{\frac{1}{3}\sim\frac{1}{2}}$, then to x far downstream, according to the classical boundary layer theory. The axial development of the wall temperature changes rather drastically depending on the value of n. When n is 0.5, the wall temperature near the inlet rises faster than the uniform porosity case, while the downstream wall temperature is suppressed. According to the correlation (61), the porosity for the case of n = 0.5 varies from 1 to 0.7 from the inlet to the exit based on the average porosity of 0.8, which changes in a moderate gradient but results in a remarkable temperature distribution variation over the axial distance. For the case of n = 1, the wall temperature was kept nearly at constant temperature of reasonably low level over most of the channel in axial direction of the channel. Thus, the results indicate that a uniform temperature can be achieved if the porosity is set to decrease linearly from the inlet to the outlet.

Furthermore, for the case of n = 1.5, the wall temperature steeply increases from the inlet and then decreases toward the exit after attaining its maximum value. The rather drastic increase in the effective thermal conductivity toward the exit explains the high sensitivity of the porosity gradient to the wall temperature distribution. Even a small increase in the metal volume fraction will result in a substantial increase in the

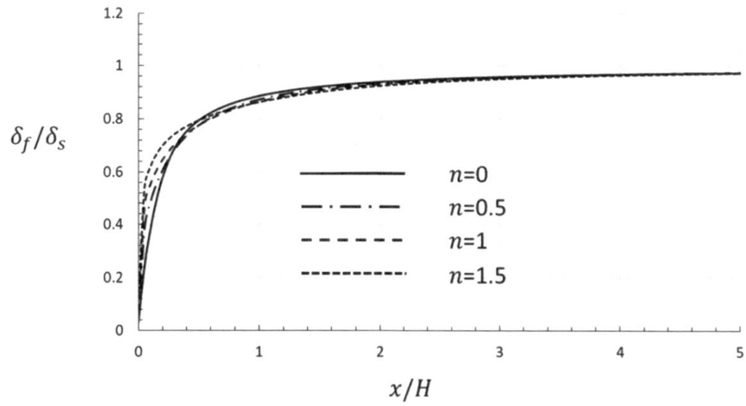

Figure 8.
Axial development of thermal boundary layer thickness ratio.

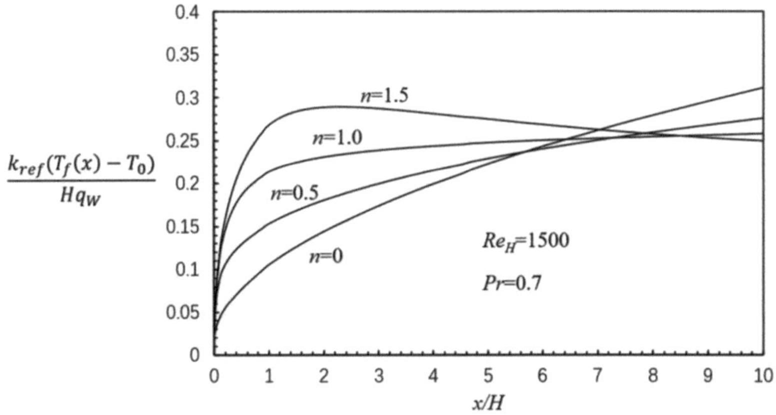

Figure 9.
Axial development of wall temperature.

effective thermal conductivity because the conductivity ratio of the metal to the fluid was set too large. Thus, the heat conducts away from the wall to the core region in the channel, keeping the wall temperature comparatively lower there.

3.4 Conclusions

In this section, the velocity profile and wall temperature development for the channel filled with axially graded metal foam were investigated analytically and numerically. The variation of the porosity follows power function of axial coordination with an exponent n based on an equal solid volume fraction. Velocity profiles for different cases were obtained based on the Brinkman-Forchheimer extended Darcy model. The solutions indicate that there exists a weak secondary flow toward the heated wall of the channel, enhancing cooling of the wall. It is found that the wall temperature is very sensitive to the axial gradient of the local porosity, even a moderate degree of the axial gradient with n = 0.5 changes the wall temperature distribution rather drastically. As a result, the wall temperature becomes almost uniform throughout the channel when n = 1.0, while, for n = 1.5, the wall temperature rises from the inlet, attains its maximum, and then decreases toward the exit. Thus, axially graded porous materials can be quite useful for designing effective cooling systems and controlling the desired working temperature and its uniformity.

4. Application of graded porous media in aeroengine cooling

The afterburner is an important component, especially in military aeroengine. It substantially improves the transient thrust of the aeroengine. However, the gas temperature at the afterburning state rises up to 2200 K, which is far higher than the temperature that the outer wall material can stand. Therefore, an efficient cooling system is definitely needed. The afterburner heat shield is often used to protect the outer wall from thermal load. However, the cooling area of the heat shield is large, film cooling holes are too many, but the coolant available for the cooling of the heat shield is relatively less. Therefore, the effective cooling of the heat shield is difficult, the high efficient cooling structure for the heat shield is definitely needed. However, due to the non-uniform pressure difference of the coolant and the mainstream, over the surface of the heat shield, the distribution of the coolant over the heat shield is also not uniform, which may lead to the thermal stress concentration on the heat shield itself. Therefore, the graded porous media may be adopted in the afterburner heat shield structure to improve its cooling effectiveness and temperature uniformity.

For convenience, in this section, four cases of impingement/effusion cooling structures embedded with uniform and non-uniform distribution pin fin array structures were considered as models to study the effects of graded structure on its heat transfer performance and cooling effectiveness. With several rib turbulator shapes, Rhee et al. [22] studied the motion and local heat/mass transfer parameters. They discovered that the location of the rib turbulators and the impingement/effusion holes affected the level of local heat/mass transfer, which was found to be boosted by the rib turbulators.

The cooling processes can be considerably impacted by lattice frame structures, which lead to higher levels of airflow disruption and an increase in the intensity of heat transfer within the flow channel thanks to the enlargement of the heat transfer surface area. Pins were shown by Funazaki [23] to be essential in improving the overall heat transfer of double-layer cooling. Li and Mongia [24] looked into how a double-walled cooling system affected heat transfer. Researchers suggested that for double wall cooling with a thin wall or thick wall, the addition of pins results in an increase of about 0.03 in overall cooling performance.

4.1 Calculation model

In this section, four cases of impingement/effusion cooling double-layer structures embedded with uniform or non-uniform (graded) distributed pin fin structures were treated as calculation models, as can be seen in **Figure 10**. Moreover, the relative location of impingement holes, effusion holes, and pin fins were also indicated.

It is significant to note that the shape of the hollow elements, as well as their diameter, remained constant. The location of the holes for both the upper and lower plates, as well as their diameters and angle of inclination, are also unchanged. Periodical boundary conditions are imposed on the side parts of the computational model. The CFD software was employed to create an unstructured mesh of the computational area, and a depiction of intricacy of the mesh is presented in **Table 1**.

Table 1 shows the main parameters of the simulation, and **Table 2** illustrates the number of grid cells for each model.

4.2 Flow structure and temperature distribution

Figure 11a–d demonstrates the characteristic features of the flow structure for each model.

In comparison to **Figure 11a** and **b**, it can be noticed that the most striking manifestations of vortices are revealed in the model whose structural elements are slightly deviated from a uniform distribution (**Figure 11b**). The difference between cases 3 and 4 (**Figure 11c** and **d**) is also noticeable. It is tangible that in the region of the lowest density of solid elements, the flow structure swirls throughout the volume of the occupied part of the space. However, otherwise, when the distribution density is high, there is a restriction of the free zone of flow propagation. In this case, good

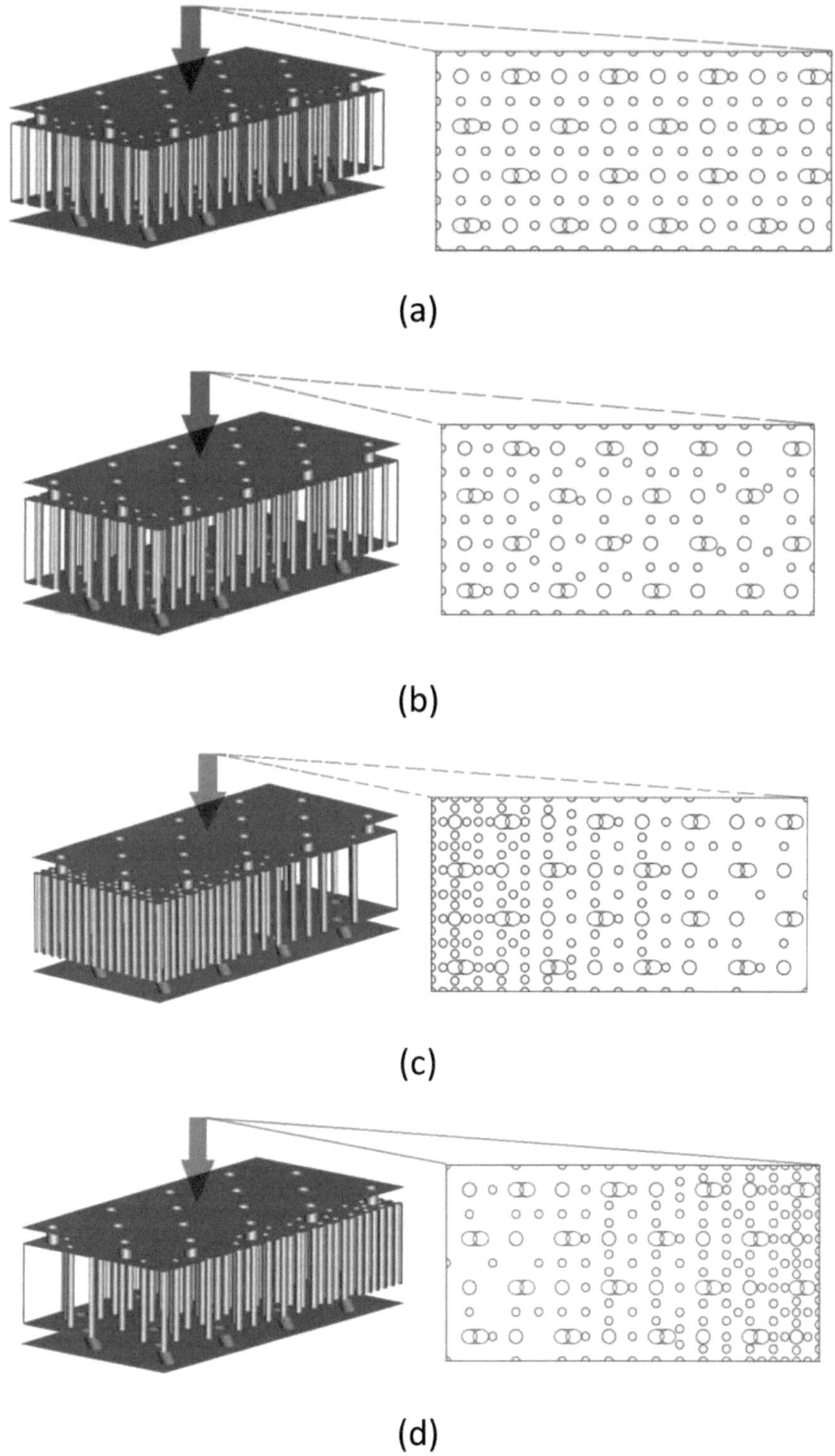

Figure 10.
Calculation models: (a) case 1 (uniform distribution; 121 pins); (b) case 2 (almost uniform distribution; 96 pins); (c) case 3 (non-uniform distribution from highest density to lowest; 166 pins); (d) case 4 (non-uniform distribution from lowest density to highest; 166 pins).

Turbulence model	SST k-ω
Mesh type	Three-dimensional
Mesh form	Orthogonal

Table 1.
The main parameters of the simulation.

Name of a case	Number of cells
Case 1 (uniform distribution)	1,809,286
Case 2 (almost uniform distribution)	1,543,251
Case 3 (non-uniform distribution from highest density to lowest)	2,304,339
Case 4 (non-uniform distribution from lowest density to highest)	2,294,669

Table 2.
The number of cells for each case.

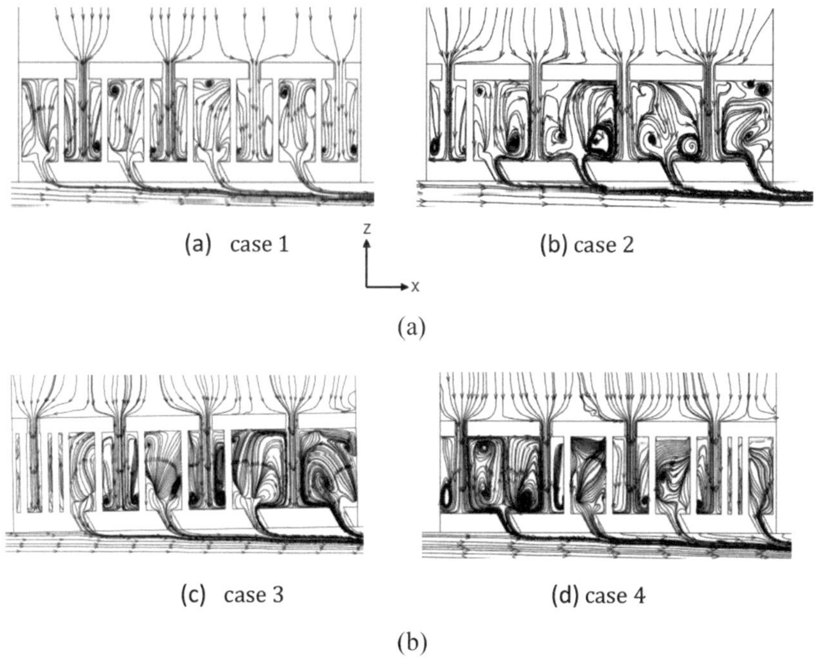

Figure 11.
Flow structures regarding the 4 cases with lattice frame elements of different distribution densities: (a) case 1, (b) case 2, (c) case 3 and (d) case 4.

mixing of the flow is provided, which is accompanied by small vortex structures throughout the volume of the area that is occupied.

The evaluation of the flow behavior analysis is very necessary, as it affects further heat exchange processes. So, **Figure 12** shows the temperature distributions for these 4 models.

Going deeper into the analysis of the flow temperature inside the model in the interlayer space, it is noted that the part of the computational model where there is a

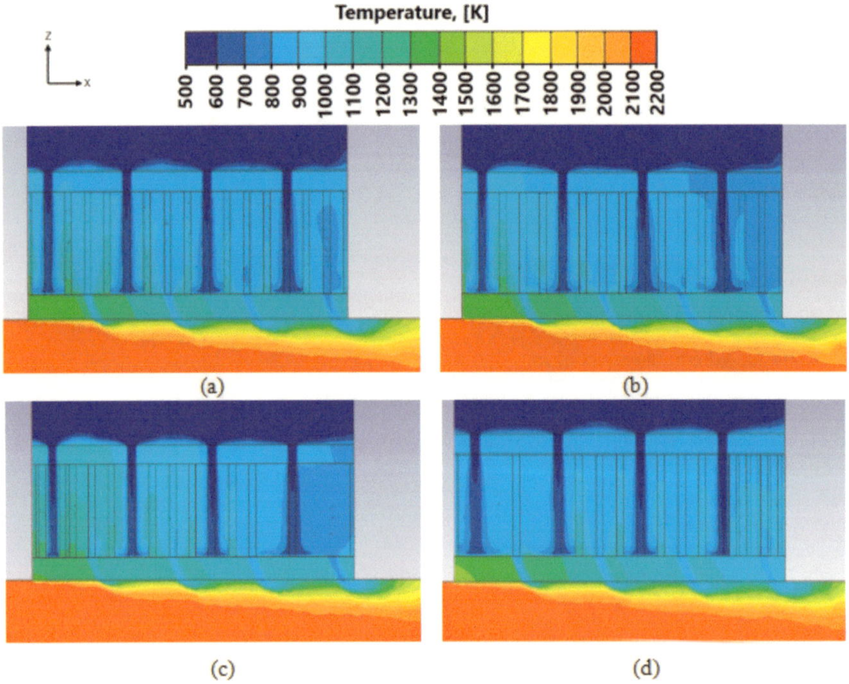

Figure 12.
Temperature distribution regarding the different cases with different distribution densities of lattice elements: (a) case 1, (b) case 2, (c) case 3, and (d) case 4.

high density of the distribution of elements of the lattice structure leads to a high intensity of heat exchange. This is due to the fact that with an increase in the distribution density, the interlayer space for mixing air flows decreases.

4.3 Overall cooling effectiveness

It is important to note that the overall cooling effectiveness is influenced by the interplay between different cooling mechanisms and their respective effectiveness values, and it is crucial to optimize the design and configuration of each cooling mechanism to achieve high overall cooling effectiveness. **Figure 13** demonstrates the distribution of the overall cooling effectiveness on 3-D models. It is vital to note that the range of values ranges from 0.3 to 0.8 according to the indicator scale above with intervals of 0.325.

From the scientific point of view, it can be noticed that with a decrease in the number of pins, the intensification of heat exchange decreases, while an increase in the same elements is accompanied by an increase in heat exchange processes. That is, the density of the distribution of lattice elements is directly proportional to the overall cooling effectiveness. This is the conclusion that can be traced from **Figure 14**.

However, to assess the overall cooling effectiveness, a graph was also constructed, presented in **Figure 14**. The blue line characterizes model 3, that is, the one whose intensification was maximal already in the first segment of the x/d ratio and whose overall cooling effectiveness is 0.515. It is revealed that the decrease in the pin distribution density is directly proportional to the decrease in the overall cooling effectiveness.

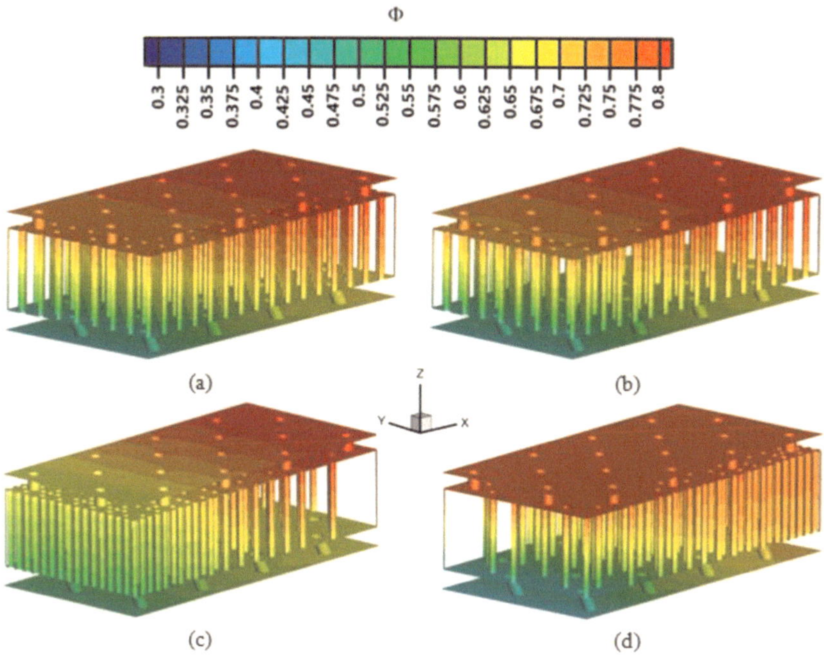

Figure 13.
3-D distribution of the overall cooling effectiveness for four models with different distribution densities of lattice elements: (a) case 1; (b) case 2; (c) case 3; (d) case 4.

The worst initial data was shown by model 4 (black line), where there was minimum distribution density at the first stage (0.47). The overall cooling effectiveness across the entire model ΔΦ in this case was 12%. Nevertheless, the black line at its minimum value began to grow sharply and reached a maximum of 0.71, while the difference in overall cooling effectiveness was as much as 25%. Such an effective indicator was achieved thanks to a rise in pins in the model under study, which only provoked enhancement in heat exchange intensification.

Summing up, the best indicators were identified in the case 3. Also, it boils down to the fact that it is necessary to introduce such lattice frame structures in which the tendency of the distribution of pins from a constructive point of view will be accompanied by an increase in the density of their distribution.

4.4 Conclusions

The simulation work of the graded lattice structures with different arrangements of pins was numerically investigated using the appropriate software (CFD). Four different cases with various distributions of the ligament elements were assumed, as well as the characteristics of the mesh construction were presented. The evaluation information showed that additional indicators that improve heat transfer processes are directly affected by the distribution density. In addition, a decrease in the distribution density leads to a decline in the overall cooling effectiveness. It is proved that when designing a model with a high density of the distribution of elements at the initial stage, there is a sharp increase in the efficiency of heat

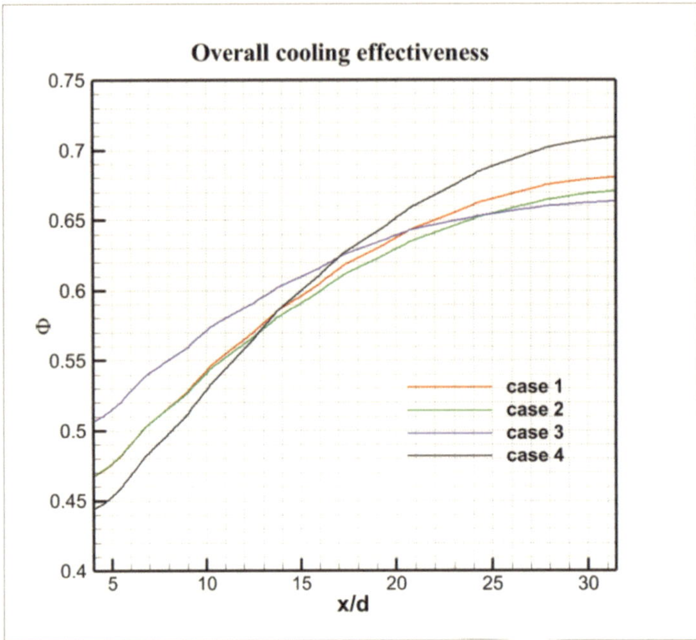

Figure 14.
The overall cooling effectiveness for four models with different distribution densities of lattice elements: red line—case 1; green line—case 2; blue line—case 3; black line—case 4.

exchange. Thus, functionally graded lattice structures can be used to achieve significantly high overall cooling effectiveness while at the same time having a high distribution density of lattice structure elements.

5. Conclusions

In this section, the flow and heat transfer in channels filled with vertical graded porous media and longitudinal graded porous media have been analytically investigated, respectively. The heat transfer characteristics and wall temperature variation principles for both cases have been presented. According to the analytical results, the vertical graded porous media in a channel enables to change the local velocity in the regions of near the wall or in the core, prompts the flow field in the low porosity region more distorted while accelerates the flow filed in the high porosity region. The highest heat transfer coefficient is obtained in the case of where the porosity decreases toward the heated wall, namely, more metal near the wall and less metal in the core region. For the case of a channel filled with axially graded porous foam structure, which follows a power function of axial distance with an exponent n, however, the velocity field is found that exist a weak secondary flow toward the heated wall, enhancing cooling of the wall. It is found that the exponent n controlling the axial gradient of the local porosity has a strong influence on the axial development of the wall temperature. A uniform wall temperature can be achieved in the case of $n = 1.0$, while a maximum temperature can be attained at the middle of the wall in the case of $n = 1.5$, instead of the fact that temperature keeps rising from the inlet to the end under the constant heat flux, which all the other researchers hold. The finding can be useful for designing effective cooling system, controlling the desired working temperature, and its uniformity.

On the other hand, the possibility of application of graded porous media in aeroengine cooling has been discussed. The numerical simulation work on the cooling of the afterburner heat shield with graded pin fin structures has been conducted. Four different cases of double-layer heat shields with various distributions of the ligament elements were constructed and examined their flow structure and cooling effectiveness. As a result, the decrease in the distribution density leads to decline in the overall cooling effectiveness, whereas a high density of pin fin elements arranged at the initial stage prompts the heat exchange and improves the overall cooling effectiveness upstream of the heat shield, although sacrifices part of that at the downstream, the gradient arrangement of pin fins improves the temperature uniformity over the surface of the heat shield. Thus, functionally graded lattice structures can be used to achieve significantly high overall cooling effectiveness while at the same time having a high distribution density of lattice structure elements.

Acknowledgements

The authors acknowledge gratefully the support from the National Science and Technology Major Project (J2019-III-0019-0063), the Innovation Capacity Support Plan in Shaanxi Province of China (Grant No.2023-CX-TD-19), the Grant of National Science Foundation of China (No.52006179), and the Fundamental Research Funds for the Central Universities (No.501XTCX2023146001).

Nomenclature

h	heat transfer coefficient
H	half channel height
k	thermal conductivity
k_e	effective stagnant thermal conductivity
K	local permeability
Nu_H	Nusselt number $Nu_H = \frac{hH}{k_e(\varepsilon_B)}$
p	pressure
q_w	wall heat flux
T	temperature
T^*	dimensionless temperature $T^* = (T - T_w)k_e(\varepsilon_B)/Hq_w$
u	Darcian velocity (apparent velocity) in x direction
u^*	dimensionless velocity $u^* = u/\left(-\frac{H^2}{\mu}\frac{dp}{dx}\right)$
u_B	bulk mean velocity
x	coordinate in the flow direction
y	coordinate normal to the channel wall
ε	porosity
ε^*	effective porosity
ε_B	bulk mean porosity
η	dimensionless vertical coordinate $\eta = y/H$
μ	viscosity

Subscripts

B	bulk mean
c	channel center-line
f	fluid phase, channel without a metal foam matrix
s	solid phase
w	wall

Author details

Xiaohui Bai[1], Cunliang Liu[1*] and Akira Nakayama[1,2]

1 School of Power and Energy, Northwestern Polytechnical University, Xi'an, China

2 Faculty of Engineering, Shizuoka University, Hamamatsu, Japan

*Address all correspondence to: liucunliang@nwpu.edu.cn

References

[1] Kamiuto K, Yee SS. Heat transfer correlations for open-cellular porous materials. International Communication and Heat Mass Transfer. 2005;**32**:947-953

[2] Dukhan N. Metal Foams: Fundamentals and Applications. Lancaster, Pennsylvania, USA: DEStech Punlications, Inc; 2013

[3] Kaviany M. Laminar flow through a porous channel bounded by isothermal parallel plates. International Journal of Heat and Mass Transfer. 1985;**28**:841-858

[4] Nakayama A, Koyama H, Kuwahara F. An analysis on forced convection in a channel filled with a Brinkman-Darcy porous medium: Exact and approximate solutions. Warme-und Stoffubertragung. 1988;**23**:291-295

[5] Calmidi VV, Mahajan RL. Forced convection in high porosity metal foams. ASME Transactions on Journal of Heat Transfer. 2000;**122**:557-565

[6] Koizumi M. FGM activities in Japan. Composites Part B: Engineering. 1997; **28**(1-2):1-4

[7] El-Wazery MS, El-Desouky AR. A review on functionally graded ceramic-metal materials. Journal of Materials and Environmental Science. 2017;**6**(5):1369-1376

[8] Wang B, Hong Y, Hou X, et al. Numerical configuration design and investigation of heat transfer enhancement in pipes filled with gradient porous materials. Energy Conversation and Management. 2015;**105**:206-215

[9] Zheng ZJ, Li MJ, He YL. Optimization of porous insert configurations for heat transfer enhancement in tubes based on genetic algorithm and CFD. International Journal of Heat and Mass Transfer. 2015;**87**:376-379

[10] Bai X, Kuwahara F, Mobedi M, Nakayama A. Forced convective heat transfer in a channel filled with a functionally graded metal foam matrix. Journal of Heat Transfer. 2018;**140**(11): 111702

[11] Iasiello M, Bianco N, Chiu WKS, Naso V. The effects of variable porosity and cell size on the thermal performance of functionally-graded foams. International Journal of Thermal Sciences. 2021;**160**:106696

[12] Mauro GM, Iasiello M, Bianco N, Chiu WKS, Naso V. Mono- and multi-objective CFD optimization of graded foam-filled channels. Materials. 2022;**15**: 968

[13] Bai X, Zheng Z, Liu C, Nakayama A. Achievement of wall temperature uniformity by axially graded porous materials. International Journal of Heat and Mass Transfer. 2022;**197**:123335

[14] Xu ZG, Zhao CY. Experimental study on pool boiling heat transfer in gradient metal foams. International Journal of Heat and Mass Transfer. 2015; **85**:824-829

[15] Yang J, Yang L, Xu C, Du X. Numerical analysis on thermal behavior of solid-liquid phase change within copper foam with varying porosity. International Journal of Heat and Mass Transfer. 2015;**84**:1008-1018

[16] Nakayama A, Kuwahara F, Kodama Y. An equation for thermal dispersion flux transport and its mathematical modelling for heat and fluid flow in a porous medium. Journal of Fluid Mechanics. 2006;**563**(25):81-96

[17] Zhang W, Li W, Nakayama A. An analytical consideration of steady-state forced convection within a nanofluid-saturated metal foam. Journal of Fluid Mechanics. 2015;**2015**(769):590-620

[18] Calmidi VV. Transport Phenomena in High Porosity Metal Foams. Denver, CO: University of Colorado; 1998

[19] Amiri A, Vafai K. Analysis of dispersion effects and non-thermal equilibrium, non-Darcian, variable porosity incompressible flow through porous media. International Journal Heat and Mass Transfer. 1994;**37**:939-954

[20] Yi Y, Bai X, Kuwahara F, Nakayama A. Analytical and numerical study on thermally developing forced convective flow in a channel filled with a highly porous medium under local thermal non-equilibrium. Transport in Porous Media. 2021;**136**:541-567

[21] Wong P, Koplik J, Tomanic JP. Conductivity and permeability of rocks. Physical Review B. 1984;**30**:6606-6614

[22] Rhee DH, Nam YW, Cho HH. Local heat/mass transfer with various rib arrangements in impingement/effusion cooling system with crossflow. Journal of Turbomachinery. 2004;**126**(4):615-626

[23] Funazaki K, Tarukawa Y, Kudo T, Matsuno S, Imai R, Yamawaki S. Heat transfer characteristics of an integrated cooling configuration of ultra-high temperature turbine blades: Experimental and numerical investigations. ASME. Paper, GT2001-0148. 2001

[24] Li S, Mongia H. An improved method for correlation of film cooling effectiveness of gas turbine combustor liners. In: 37th Joint Propulsion Conference and Exhibit, AIAA; July, 2001; Salt Lake City, Utah, USA. p. 3268

Chapter 8

Dissolution of Porous Media and Underground Cavities: A Multi-Scale View

Jianwei Guo

Abstract

To better understand the dissolution of porous media and underground cavities is very important in various applications. In this chapter, pore-scale dissolution model, which involves thermodynamic equilibrium or nonlinear reactive boundary conditions, is upscaled into Darcy-scale using the method of volume averaging. In the Darcy-scale model, several effective parameters are employed to describe the average behaviors of the pore-scale features, and they can be obtained by solving specific closure problems. The developed Darcy-scale model is validated by taking the dissolution of a gypsum pillar as an example. The results show that when Péclet and Reynolds number are within the assumptions to apply volume averaging, computation results using Darcy-scale model agree very well with direct numerical simulations. However, when they go beyond certain limits, 3D effects have to be taken into consideration.

Keywords: dissolution, porous media, cavity, upscaling, effective parameters

1. Introduction

Dissolution of underground cavities and porous media are encountered in various subsurface processes, such as the evolution of karst formations [1], solutional salt mining [2], the geological sequestration of CO_2 [3, 4], the improved oil recovery [5–8], and so on. Many challenges we face in such problems relate to the crucial need of better description of reactive transport. Generally speaking, the evolution of underground cavities involves multiple scales. **Figure 1** schematically shows the various scales associated to the injection of a dissolving fluid in a geological site (e.g., acid injection in oil recovery and solution mining of salt). On the one hand, it is possible to have the dissolution of a porous formation, which may lead to a porous dissolved region before the formation of a true fluid cavity. On the other hand, it may be the dissolution of an impervious rock formation with a more direct route to a fluid cavity. The heterogeneities generated during the process will make the multi-scale feature even more complicated [9].

It is a common sense that pore-scale model is the most accurate way to get deep insight into dissolution processes, in which the chemical-physical interactions take place and the full physics of dissolution is considered [10–13]. In pore-scale modeling,

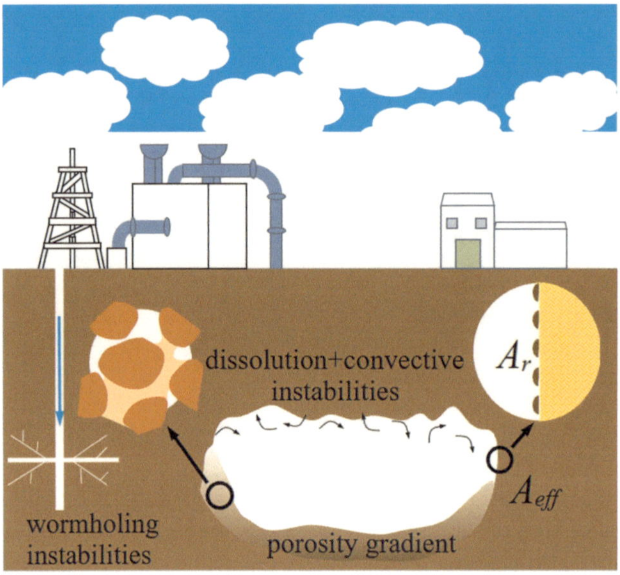

Figure 1.
Schematic illustration of the multiple scales encountered in the dissolution of underground cavities.

the governing equations are solved with specific boundary conditions at the receding interfaces, without any assumption about the geometric structures of the porous media. And different phases are separated by the sharp interfaces [11, 12]. Such kind of pore-scale modeling was carried out using different methods, such as the Arbitrary-Lagrangian–Eulerian (ALE) framework [11, 14], level set method [15, 16], phase-field method [17], pore network models (PNM) [18, 19], and so on. However, pore-scale modeling is numerically difficult and expensive because of several reasons. Firstly, it is often very difficult to get the small-scale information, such as the structure of the skeleton and the voids. Secondly, people are more interested in large-scale behaviors in industrial or geological applications, which may range from decimeter to meter. However, to take all the microscopic pore-scale details into consideration is impractical and even impossible. Moreover, it is also a big chanllenge to explicitly treat the moving interface with large deformations, which may be created by the dissolution process [11, 13, 20]. To overcome this difficulty, an efficient way is to apply some sort of upscaling to the pore-scale models to study the dissolution problem from a macro-scale viewpoint. The macro-/Darcy-scale models filter pore-scale details below a cut-off length while representing them by several macro-scale effective parameters instead. These effective paramters can be solved through corresponding "closure problems" generated during the upscaling process. Since the middle of the last century, there appears a large number of studies on the upscaling of transport equations for a given chemical species. For example, the early fundamental works of Taylor [21, 22] and Aris [23] dealt with *passive* dispersion in porous media, i.e., advection and diffusion, without consideration of interaction at the fluid-solid interfaces. This problem was studied later using various upscaling tools [24–26], which led to the classical macro-scale dispersion theory and the proposition of local closure problems used to calculate the dispersion tensor components for various representative unit cells. An important progress was made by extending this theory to investigate *active* dispersion [27–30], i.e., mass transport at the phase interface. Thereby, this led to the generation

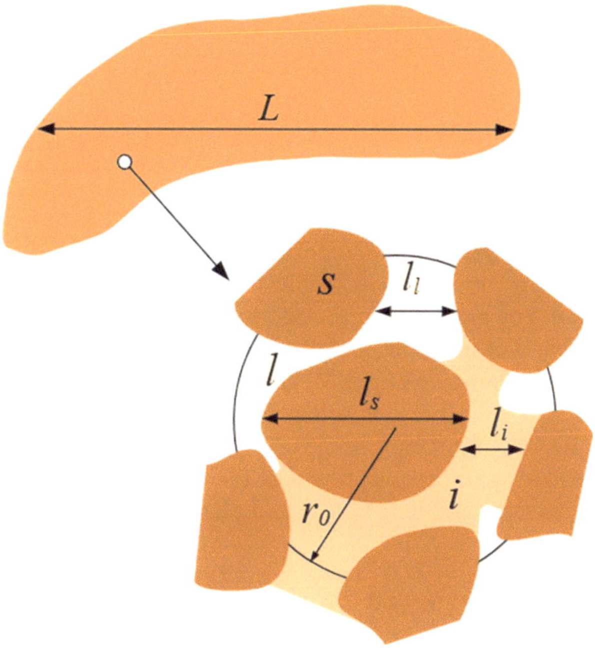

Figure 2.
Schematic illustration of the multiple scales associated with dissolution in a porous medium.

of an effective mass exchange term in the resulting Darcy-scale models, which incorporates the pore-scale thermodynamic equilibrium or reactive boundary condition at the interface [29].

In this chapter, we consider a heterogeneous system as schematically illustrated in **Figure 2**. The pore-scale characteristic lengths are denoted as ℓ_α, where $\alpha = l, i$ and s for the the liquid phase (in general water + dissolved species) flowing into the pore-space, the insoluble material, and the soluble material (which is assumed to contain only one component here), respectively. The macro-scale characteristic length is denoted as L. A third length-scale is defined as r_0, which is the radius of the representative elementary volume (REV). Upscaling can be done to develop a Darcy-scale model only when the hypothesis of scale separation expressed by

$$\ell_\alpha \ll r_0 \ll L \tag{1}$$

is fulfilled.

The rest of this chapter is organized as follows. The pore-scale model to describe a dissolution problem will be presented at first. Then, volume averaging method will be used to upscale this pore-scale model into macro-scale. Closure problems used to solve the effective parameters emerging in the macro-scale model will be introduced subsequently and examples to show the validation and robustness of the macro-scale model will be given at last.

2. Pore-scale model

In this section, the pore-scale model is developed. The focus is on the mass transport problem, leaving aside the momentum and energy eqs. A hierarchical, multi-scale system is considered as schematically depicted in **Figure 2**. The momentum balance equations in the fluid phase are written as

$$\frac{\partial \rho_l \mathbf{v}_l}{\partial t} + \rho_l \mathbf{v}_l \cdot \nabla \mathbf{v}_l = -\nabla p_l + \rho_l \mathbf{g} + \mu_l \nabla^2 \mathbf{v}_l \quad \text{in the liquid phase} \tag{2}$$

$$\mathbf{v}_l - \mathbf{n}_{ls}\mathbf{n}_{ls} \cdot \mathbf{v}_l = 0 \quad \text{at } A_{ls} \tag{3}$$

where ρ_l is the liquid density, $\mathbf{v}_l$ the pore-scale liquid velocity, $\mathbf{g}$ the gravity, μ_l the dynamic viscosity, and A_{ls} the interfacial area between the soluble solid and the liquid phase. For most underground dissolution processes, the density variation is negligible. For example, as the second most soluble rock, gypsum has a solubility in water of about 2.63 kg m^{-3}, which leads to the water density variation of less than 0.3%. However, when considering the dissolution of salt, buoyancy effects may be significant because of the large density variation. Nevertheless, considering the assumption that the liquid density is a constant, mass balance equations for the liquid phase and the dissolved species can be given by

$$\nabla \cdot \mathbf{v}_l = 0 \quad \text{in the liquid phase} \tag{4}$$

$$\frac{\partial \omega_l}{\partial t} + \mathbf{v}_l \cdot \nabla \omega_l = \nabla \cdot (D_l \nabla \omega_l) \quad \text{in the liquid phase} \tag{5}$$

with ω_l the mass fraction of the followed ion (e.g., Ca^{2+} in the context of gypsum dissolution) in the liquid phase and D_l the molecular diffusion coefficient.

When considering a reactive boundary condition at the solid-liquid interface, the following kinetic condition can be written

$$\begin{aligned} \mathbf{n}_{ls} \cdot (\rho_l \omega_l (\mathbf{v}_l - \mathbf{w}_{sl}) - \rho_l D_l \nabla \omega_l) &= -M_c k_s \left(1 - \frac{\omega_l}{\omega_{eq}}\right)^n \\ &= \mathbf{n}_{ls} \cdot (\rho_s \omega_s (\mathbf{v}_s - \mathbf{w}_{sl})) \quad \text{at} A_{ls} \end{aligned} \tag{6}$$

In this equation, ρ_s is the solid density, ω_s is the mass fraction of the followed component in the solid phase, $\mathbf{v}_s$ is the solid velocity, $\mathbf{w}_{sl}$ is the interface velocity, M_c is the molar weight of the followed ion, k_s is the reaction rate coefficient, and n is the nonlinear reaction order. The following boundary condition can be written from the mass balance of the solid phase

$$\begin{aligned} \mathbf{n}_{ls} \cdot (\rho_s (\mathbf{v}_s - \mathbf{w}_{sl})) &= \nu_s \mathbf{n}_{ls} \cdot (\rho_l \omega_l (\mathbf{v}_l - \mathbf{w}_{sl}) - \rho_l D_l \nabla \omega_l) \\ &= -\nu_s M_c k_s \left(1 - \frac{\omega_l}{\omega_{eq}}\right)^n \quad \text{at} A_{ls} \end{aligned} \tag{7}$$

where $\nu_s = M_s / M_c$ and M_s is the molar weight of solid phase. The total mass balance boundary condition writes

$$\mathbf{n}_{ls} \cdot (\rho_l (\mathbf{v}_l - \mathbf{w}_{sl})) = \mathbf{n}_{ls} \cdot (\rho_s (\mathbf{v}_s - \mathbf{w}_{sl})) \quad \text{at} A_{ls} \tag{8}$$

In this chapter, the solid phase is assumed not moving, i.e., $\mathbf{v}_s = 0$, thus $\mathbf{v}_s$ can be discarded in the boundary conditions. Consequently, Eqs. (7) and (8) can be transformed into

$$\mathbf{n}_{ls}\cdot\mathbf{w}_{sl}=\rho_s^{-1}M_sk_s\left(1-\frac{\omega_l}{\omega_{eq}}\right)^n \qquad \text{at} A_{ls} \tag{9}$$

to calculate the interface velocity, and

$$\mathbf{n}_{ls}\cdot\rho_l\mathbf{v}_l=\mathbf{n}_{ls}\cdot(\rho_l-\rho_s)\mathbf{w}_{sl}=-\left(1-\frac{\rho_l}{\rho_s}\right)M_sk_s\left(1-\frac{\omega_l}{\omega_{eq}}\right)^n \text{at} A_{ls} \tag{10}$$

to get the liquid velocity, respectively. Eqs. (7) and (8) also lead to

$$\mathbf{n}_{ls}\cdot\mathbf{w}_{sl}=\frac{\nu_s}{\rho_s(1-\nu_s\omega_l)}\mathbf{n}_{ls}\cdot(\rho_lD_l\nabla\omega_l) \qquad \text{at} A_{ls} \tag{11}$$

in the case of an equilibrium condition, which provides a more convenient way to calculate the interface velocity. For gypsum dissolution, $\mathbf{w}_{sl}$ is in the order of 10^{-7}m s^{-1} and it is even smaller for the dissolution of carbonate rocks, which is usually negligible compared to the liquid velocity. Considering also Eq. (3), Eq. (6) can be simplified into

$$\mathbf{n}_{ls}\cdot(-\rho_lD_l\nabla\omega_l)\approx-M_ck_s\left(1-\frac{\omega_l}{\omega_{eq}}\right)^n \quad \text{at } A_{ls} \tag{12}$$

Collect the above mass balance equations and boundary conditions, the pore-scale problem can bc written as

$$\frac{\partial\rho_l}{\partial t}+\nabla\cdot(\rho_l\mathbf{v}_l)=0 \qquad \text{in the liquid phase} \tag{13}$$

$$\frac{\partial\rho_l\omega_l}{\partial t}+\nabla\cdot(\rho_l\omega_l\mathbf{v}_l)=\nabla\cdot(\rho_lD_l\nabla\omega_l) \qquad \text{in the liquid phase} \tag{14}$$

$$\mathbf{n}_{ls}\cdot(\rho_l\omega_l(\mathbf{v}_l-\mathbf{w}_{sl})-\rho_lD_l\nabla\omega_l)=-M_ck_s\left(1-\frac{\omega_l}{\omega_{eq}}\right)^n \text{at} A_{ls} \tag{15}$$

$$\begin{aligned}\mathbf{n}_{ls}\cdot(-\rho_s\mathbf{w}_{sl})&=\mathbf{n}_{ls}\cdot(\rho_l(\mathbf{v}_l-\mathbf{w}_{sl}))\\&=\nu_s\mathbf{n}_{ls}\cdot(\rho_l\omega_l(\mathbf{v}_l-\mathbf{w}_{sl})-\rho_lD_l\nabla\omega_l)\\&=-M_sk_s\left(1-\frac{\omega_l}{\omega_{eq}}\right)^n \qquad \text{at} A_{ls}\end{aligned} \tag{16}$$

$$\frac{\partial\rho_s}{\partial t}=0 \quad \text{in the solid phase} \tag{17}$$

$$\mathbf{n}_{li}\cdot(\rho_l\omega_l\mathbf{v}_l-\rho_lD_l\nabla\omega_l)=0 \qquad \text{at } A_{li} \tag{18}$$

$$\mathbf{n}_{li}\cdot\mathbf{v}_l=0 \qquad \text{at } A_{li} \tag{19}$$

Here, A_{li} denotes the interfacial area between the insoluble solid and the liquid phase.

This pore-scale model should be completed with the momentum balance equations, such as Eqs. (2) and (3). Moreover, the momentum balance equations are independent of the mass balance equations, because the characteristic time for the relaxation of viscous flow is much shorter than the dissolution characteristic time, and the liquid velocity and viscosity are assumed constant. Thus, the focus of this chapter is on the mass transport problem.

Finally, when the boundary condition is thermodynamic equilibrium instead of the reactive condition discussed above, Eq. (15) should be replaced by

$$\omega_l = \omega_{eq} \qquad \text{at } A_{ls} \tag{20}$$

3. Darcy-scale model

For the dissolution taking place in a porous medium such as illustrated in **Figure 1**, it is often not practical to take into account of all the pore-scale details by direct numerical modeling in a L-scale problem, even if this is the more secure way of handling the geometry evolution [10]. Consequently, attempts have been made to filter the pore-scale information below a cutoff length through upscaling techniques. In the past decades, several upscaling methods have been developed, such as volume averaging [31], ensemble averaging [32, 33], moments matching [34, 35], and multi-scale asymptotic [36]. In this section, the macro-scale model is developed using the technique of volume averaging, and this present work extends the contribution of Refs. [30, 37, 38].

3.1 Averages and averaged equations

The superficial and intrinsic average of the pore-scale liquid velocity, i.e., **V** and $\mathbf{U}_l$, are defined as

$$\mathbf{V}_l = \langle \mathbf{v}_l \rangle = \frac{1}{V}\int_{V_l} \mathbf{v}_l dV \tag{21}$$

and

$$\mathbf{U}_l = \frac{1}{V_l}\int_{V_l} \mathbf{v}_l dV = \frac{\mathbf{V}_l}{\varepsilon_l} \tag{22}$$

respectively, where V denotes the volume of the representative unit cell as illustrated in **Figure 2**, V_l is the volume of the liquid phase within it, and ε_l is the porosity given by

$$\varepsilon_l = \frac{1}{V}\int_{V_l} dV \tag{23}$$

The intrinsic average of the mass fraction of the followed species in the liquid phase is written as

$$\Omega_l = \langle \omega_l \rangle^l = \frac{1}{V_l}\int_{V_l} \omega_l dV \tag{24}$$

The pore-scale and macro-scale variables are related through deviations by

$$\omega_l = \Omega_l + \tilde{\omega}_l \tag{25}$$

$$\mathbf{v}_l = \mathbf{U}_l + \tilde{\mathbf{v}}_l \tag{26}$$

with $\tilde{\omega}_l$ and $\tilde{\mathbf{v}}_l$ representing the deviation of mass fraction of the followed component and the velocity of the liquid phase, respectively.

Averaging Eq. (13) leads to

$$\frac{\partial \varepsilon_l \rho_l}{\partial t} + \nabla \cdot (\rho_l \mathbf{V}_l) = -K_s \tag{27}$$

Here, the mass exchange term is given by

$$K_s = \frac{1}{V} \int_{A_{ls}} \mathbf{n}_{ls} \cdot \rho_l (\mathbf{v}_l - \mathbf{w}_{sl}) dA \tag{28}$$

Applying average to the solid mass balance equation gives

$$\frac{\partial \varepsilon_s \rho_s}{\partial t} = K_s \tag{29}$$

where Eq. (16) has been used to obtain

$$\begin{aligned} K_s &= \frac{1}{V} \int_{A_{ls}} \mathbf{n}_{ls} \cdot (-\rho_s \mathbf{w}_{sl}) dA = \frac{1}{V} \int_{A_{ls}} \mathbf{n}_{ls} \cdot \rho_l (\mathbf{v}_l - \mathbf{w}_{sl}) dA \\ &= -a_{vl} M_s k_s \left\langle \left(1 - \frac{\omega_l}{\omega_{eq}}\right)^n \right\rangle_{ls} \end{aligned} \tag{30}$$

with the surface average and specific area defined as

$$\langle \star \rangle_{ls} = \frac{1}{A_{ls}} \int_{A_{ls}} \star dA \tag{31}$$

and

$$a_{vl} = \frac{1}{V} \int_{A_{ls}} dA \tag{32}$$

respectively.

Finally, the mass balance equation for the followed ion writes

$$\begin{aligned} &\frac{\partial \varepsilon_l \rho_l \Omega_l}{\partial t} + \nabla \cdot (\varepsilon_l \rho_l \Omega_l \mathbf{U}_l) + \underbrace{\nabla \cdot (\rho_l \langle \tilde{\omega}_l \tilde{\mathbf{v}}_l \rangle)}_{\text{dispersion}} = \\ &\nabla \cdot \left(\varepsilon_l \rho_l D_l \nabla \Omega_l + \underbrace{\frac{1}{V} \int_{A_{ls}} \mathbf{n}_{ls} \rho_l D_l \tilde{\omega}_l dA + \frac{1}{V} \int_{A_{li}} \mathbf{n}_{li} \rho_l D_l \tilde{\omega}_l dA}_{\text{tortuosity}} \right) - K_c \end{aligned} \tag{33}$$

The total mass exchange term K_c is given by

$$K_c = \frac{M_c}{M_s} K_s = -a_{vl} M_c k_s \left\langle \left(1 - \frac{\omega_l}{\omega_{eq}}\right)^n \right\rangle_{ls} \tag{34}$$

Using Eq. (27), Eq. (33) can be written as

$$\begin{aligned} &\varepsilon_l \rho_l \frac{\partial \Omega_l}{\partial t} + \varepsilon_l \rho_l \mathbf{U}_l \cdot \nabla \Omega_l + \underbrace{\nabla \cdot (\rho_l \langle \tilde{\omega}_l \tilde{\mathbf{v}}_l \rangle)}_{\text{dispersion}} = \\ &\nabla \cdot \left(\varepsilon_l \rho_l D_l \nabla \Omega_l + \underbrace{\frac{1}{V} \int_{A_{ls}} \mathbf{n}_{ls} \rho_l D_l \tilde{\omega}_l dA + \frac{1}{V} \int_{A_{li}} \mathbf{n}_{li} \rho_l D_l \tilde{\omega}_l dA}_{\text{tortuosity}} \right) - K_c + K_s \Omega_l \end{aligned} \tag{35}$$

3.2 Deviation equations and closure problems

Up to now, the equations are still macro- and micro-scale coupled, and we need to find out an estimation for the concentration deviations. This is done through the following steps. Firstly, substituting Eq. (25) into the pore-scale mass balance equations. Secondly, dividing the averaged equation (Eq. (35)) by ε_l and thirdly subtracting it from the pore-scale mass balance equations. It is noteworthy that the spatial variations of the volume fraction and liquid density over the REV are neglected. Using the following equations

$$\nabla \cdot (\rho_l \omega_l \mathbf{v}_l) = \rho_l \omega_l \nabla \cdot \mathbf{v}_l + \rho_l \mathbf{v}_l \cdot \nabla \omega_l = \rho_l \mathbf{v}_l \cdot \nabla \Omega_l + \rho_l \mathbf{v}_l \cdot \nabla \tilde{\omega}_l \tag{36}$$

$$\begin{aligned} &-\varepsilon_l^{-1} \nabla \cdot \left(\varepsilon_l \rho_l D_l \nabla \Omega_l + \frac{1}{V} \int_{A_{ls}} \mathbf{n}_{ls} \rho_l D_l \tilde{\omega}_l dA + \frac{1}{V} \int_{A_{li}} \mathbf{n}_{li} \rho_l D_l \tilde{\omega}_l dA \right) \approx \\ &-\nabla \cdot \left(\rho_l D_l \nabla \Omega_l + \frac{\varepsilon_l^{-1}}{V} \int_{A_{ls}} \mathbf{n}_{ls} \rho_l D_l \tilde{\omega}_l dA + \frac{\varepsilon_l^{-1}}{V} \int_{A_{li}} \mathbf{n}_{li} \rho_l D_l \tilde{\omega}_l dA \right) \end{aligned} \tag{37}$$

$$\begin{aligned} \varepsilon_l^{-1} K_c - \varepsilon_l^{-1} K_s \Omega_l = \varepsilon_l^{-1} \frac{1}{V} \int_{A_{ls}} \mathbf{n}_{ls} (\rho_l \tilde{\omega}_l (\mathbf{v}_l - \mathbf{w}_{sl}) - \rho_l D_l \nabla \tilde{\omega}_l) dA \\ - \rho_l D_l \varepsilon_l^{-1} \frac{1}{V} \int_{A_{ls}} \mathbf{n}_{ls} \, dA \cdot \nabla \Omega_l \end{aligned} \tag{38}$$

$$\nabla \cdot \left(\frac{\varepsilon_l^{-1}}{V} \int_{A_{ls}} \mathbf{n}_{ls} \rho_l D_l \tilde{\omega}_l dA + \frac{\varepsilon_l^{-1}}{V} \int_{A_{li}} \mathbf{n}_{li} \rho_l D_l \tilde{\omega}_l dA \right) \ll \nabla \cdot (\rho_l D_l \nabla \tilde{\omega}_l) \tag{39}$$

one obtains

$$\begin{aligned} &\rho_l \frac{\partial \tilde{\omega}_l}{\partial t} + \rho_l \mathbf{v}_l \cdot \nabla \tilde{\omega}_l + \rho_l \tilde{\mathbf{v}}_l \cdot \nabla \Omega_l - \varepsilon_l^{-1} \nabla \cdot (\rho_l \langle \tilde{\omega}_l \tilde{\mathbf{v}}_l \rangle) = \nabla \cdot (\rho_l D_l \nabla \tilde{\omega}_l) \\ &\quad + \frac{\varepsilon_l^{-1}}{V} \int_{A_{ls}} \mathbf{n}_{ls} \cdot \rho_l \tilde{\omega}_l (\mathbf{v}_l - \mathbf{w}_{sl}) dA - \frac{\varepsilon_l^{-1}}{V} \int_{A_{ls}} \mathbf{n}_{ls} \cdot \rho_l D_l \nabla \tilde{\omega}_l dA \\ &\quad - \rho_l D_l \varepsilon_l^{-1} \frac{1}{V} \int_{A_{ls}} \mathbf{n}_{ls} dA \cdot \nabla \Omega_l \end{aligned} \tag{40}$$

In addition, the assumption

$$\Omega_{ls} = \langle \omega_l \rangle_{ls} = \frac{1}{A_{ls}} \int_{A_{ls}} \omega_l dA \approx \Omega_l \tag{41}$$

is used, so any integral of the form $\frac{1}{V} \int_{A_{ls}} \omega_l dA$ will be approximated by

$$\frac{1}{V} \int_{A_{ls}} \omega_l dA \approx a_{vl} \Omega_l \tag{42}$$

Using Eq. (15), the boundary condition for the deviation can be written as

$$\begin{aligned} &\mathbf{n}_{ls} \cdot (\rho_l \Omega_l (\mathbf{v}_l - \mathbf{w}_{sl})) + \mathbf{n}_{ls} \cdot (\rho_l \tilde{\omega}_l (\mathbf{v}_l - \mathbf{w}_{sl})) + \mathbf{n}_{ls} \cdot (-\rho_l D_l \nabla \tilde{\omega}_l) \\ &\quad + \mathbf{n}_{ls} \cdot (-\rho_l D_l \nabla \Omega_l) = -M_c k_s \left(1 - \frac{\Omega_l + \tilde{\omega}_l}{\omega_{eq}}\right)^n \qquad \text{at} \, A_{ls} \end{aligned} \tag{43}$$

The crucial assumption now, which will be tested quantitatively against DNS, is to estimate the nonlinear reaction rate by a first-order Taylor expansion, i.e.,

$$\begin{aligned} &\mathbf{n}_{ls} \cdot (\rho_l \Omega_l (\mathbf{v}_l - \mathbf{w}_{sl})) + \mathbf{n}_{ls} \cdot (\rho_l \tilde{\omega}_l (\mathbf{v}_l - \mathbf{w}_{sl})) + \mathbf{n}_{ls} \cdot (-\rho_l D_l \nabla \tilde{\omega}_l) \\ &\quad + \mathbf{n}_{ls} \cdot (-\rho_l D_l \nabla \Omega_l) = -M_c k_s \left(1 - \frac{\Omega_l}{\omega_{eq}}\right)^n + \frac{n M_c k_s}{\omega_{eq}} \left(1 - \frac{\Omega_l}{\omega_{eq}}\right)^{n-1} \tilde{\omega}_l \; \text{at} \, A_{ls} \end{aligned} \tag{44}$$

or

$$\begin{aligned} &\mathbf{n}_{ls} \cdot (\rho_l \Omega_l (\mathbf{v}_l - \mathbf{w}_{sl})) + \mathbf{n}_{ls} \cdot (\rho_l \tilde{\omega}_l (\mathbf{v}_l - \mathbf{w}_{sl})) + \mathbf{n}_{ls} \cdot (-\rho_l D_l \nabla \tilde{\omega}_l) \\ &\quad + \mathbf{n}_{ls} \cdot (-\rho_l D_l \nabla \Omega_l) = -\frac{M_c k_s}{\omega_{eq}} \left(1 - \frac{\Omega_l}{\omega_{eq}}\right)^{n-1} \left(\omega_{eq} - \Omega_l - n \tilde{\omega}_l\right) \text{at} \, A_{ls} \end{aligned} \tag{45}$$

Similarly, Eq. (18) gives

$$\mathbf{n}_{li} \cdot (-\rho_l D_l \nabla \tilde{\omega}_l) + \mathbf{n}_{li} \cdot (-\rho_l D_l \nabla \Omega_l) = 0 \qquad \text{at} \, A_{li} \tag{46}$$

Following the developments described in the literature [30, 37, 38], $\left(\Omega_l - \omega_{eq}\right)$ and $\nabla \Omega_l$ appeared in the above governing equations can be considered as source terms because they can be generated from the deviation terms. The mathematical structure of the above macro- and micro-scale coupled equations enables to write an approximate solution, using the expansion of the source terms and their higher derivatives as follows

$$\tilde{\omega}_l = s_l \left(\Omega_l - \omega_{eq}\right) + \mathbf{b}_l \cdot \nabla \Omega_l + \ldots \tag{47}$$

which gives for the first-order derivatives

$$\nabla \tilde{\omega}_l = \nabla s_l \left(\Omega_l - \omega_{eq}\right) + (\nabla \mathbf{b}_l + s_l \mathbf{I}) \cdot \nabla \Omega_l + \ldots \tag{48}$$

According to the length-scale constraints $l_s, l_i, l_l \ll r_0 \ll L$, terms involving higher-order derivatives than $\nabla \Omega_l$ in Eq. (48) can be neglected. Substituting Eqs. (47) and

(48) into Eq. (40), collecting terms for $\left(\Omega_l - \omega_{eq}\right)$ and $\nabla\Omega_l$ and then employing the fundamental lemma stated in Ref. [30], two closure problems shown below were obtained, which are dependent on the macro-scale mass fraction Ω_l.

Problem I:

$$\rho_l \frac{\partial s_l}{\partial t} + \rho_l \mathbf{v}_l \cdot \nabla s_l = \nabla \cdot (\rho_l D_l \nabla s_l) + \varepsilon_l^{-1} \rho_l X_l \quad \text{in the liquid phase} \tag{49}$$

$$\begin{aligned} &\mathbf{n}_{ls} \cdot \frac{\Omega_l}{\Omega_l - \omega_{eq}} \rho_l (\mathbf{v}_l - \mathbf{w}_{sl}) + \mathbf{n}_{ls} \cdot (\rho_l s_l (\mathbf{v}_l - \mathbf{w}_{sl}) - \rho_l D_l \nabla s_l) \\ &= \frac{M_c k_s}{\omega_{eq}} \left(1 - \frac{\Omega_l}{\omega_{eq}}\right)^{n-1} (1 + n\, s_l) \qquad \text{at} A_{ls}(t) \end{aligned} \tag{50}$$

$$\mathbf{n}_{li} \cdot (-\rho_l D_l \nabla s_l) = 0 \quad \text{at} A_{li} \tag{51}$$

$$\langle s_l \rangle = 0 \quad \text{in the liquid phase} \tag{52}$$

$$s_l(\mathbf{x} + \mathbf{l}_i) = s_l(\mathbf{x}) \tag{53}$$

here, $A_{ls}(t)$ was used to indicate the geometry evolution and the following notations were defined

$$\begin{aligned} \rho_l X_l = &\frac{1}{V} \int_{A_{ls}} \mathbf{n}_{ls} \cdot \rho_l s_l (\mathbf{v}_l - \mathbf{w}_{sl}) dA - \frac{1}{V} \int_{A_{ls}} \mathbf{n}_{ls} \cdot \rho_l D_l \nabla s_l dA \\ &- \frac{1}{V} \int_{A_{li}} \mathbf{n}_{li} \cdot \rho_l D_l \nabla s_l dA \end{aligned} \tag{54}$$

Making average to Eq. (49) and considering $\rho_l \langle \mathbf{v}_l \cdot \nabla s_l \rangle = \rho_l \nabla \cdot \langle \mathbf{v}_l s_l \rangle = 0$, we can write

$$\begin{aligned} \rho_l \frac{\partial \langle s_l \rangle}{\partial t} = &- \frac{1}{V} \int_{A_{ls}} \mathbf{n}_{ls} \cdot \rho_l s_l (\mathbf{v}_l - \mathbf{w}_{sl}) dA + \frac{1}{V} \int_{A_{ls}} \mathbf{n}_{ls} \cdot \rho_l D_l \nabla s_l dA \\ &+ \frac{1}{V} \int_{A_{li}} \mathbf{n}_{li} \cdot \rho_l D_l \nabla s_l dA + \rho_l X_l \equiv 0 \end{aligned} \tag{55}$$

which confirms the consistency of the proposed closure problem.

Problem II:

$$\rho_l \frac{\partial \mathbf{b}_l}{\partial t} + \rho_l \mathbf{v}_l \cdot (\nabla \mathbf{b}_l + s_l \mathbf{I}) + \rho_l \tilde{\mathbf{v}}_l - \varepsilon_l^{-1} \rho_l \langle \tilde{\mathbf{v}}_l s_l \rangle = \nabla \cdot (\rho_l D_l (\nabla \mathbf{b}_l + s_l \mathbf{I})) + \varepsilon_l^{-1} \rho_l \mathbf{u}_l \text{ in the liquid phase} \tag{56}$$

$$\begin{aligned} &\mathbf{n}_{ls} \cdot (\rho_l (\mathbf{v}_l - \mathbf{w}_{sl}) \mathbf{b}_l - \rho_l D_l (\nabla \mathbf{b}_l + s_l \mathbf{I})) - \rho_l D_l \mathbf{n}_{ls} \\ &= n \frac{M_c k_s}{\omega_{eq}} \left(1 - \frac{\Omega_l}{\omega_{eq}}\right)^{n-1} \mathbf{b}_l \ \text{at} A_{ls}(t) \end{aligned} \tag{57}$$

$$\mathbf{n}_{li} \cdot (-\rho_l D_l (\nabla \mathbf{b}_l + s_l \mathbf{I})) = \rho_l D_l \mathbf{n}_{li} \quad \text{at} A_{li} \tag{58}$$

$$\langle \mathbf{b}_l \rangle = 0 \quad \text{in} V_l \tag{59}$$

$$\mathbf{b}_l(\mathbf{x} + \mathbf{l}_i) = \mathbf{b}_l(\mathbf{x}) \tag{60}$$

in which

$$\rho_l \mathbf{u}_l = \frac{1}{V}\int_{A_{ls}} \mathbf{n}_{ls} \cdot \rho_l(\mathbf{v}_l - \mathbf{w}_{sl})\mathbf{b}_l dA - \frac{1}{V}\int_{A_{ls}} \mathbf{n}_{ls} \cdot \rho_l D_l(\nabla \mathbf{b}_l + s_l \mathbf{I}) - \frac{1}{V}\int_{A_{ls}} \mathbf{n}_{ls}\rho_l D_l dA \tag{61}$$

Using the following relation

$$-\nabla \varepsilon_l = \frac{1}{V}\int_{A_{ls}} \mathbf{n}_{ls} dA + \frac{1}{V}\int_{A_{li}} \mathbf{n}_{li} dA = 0 \tag{62}$$

it may be written that

$$-\frac{1}{V}\int_{A_{ls}} \mathbf{n}_{ls}\rho_l D_l dA = \frac{1}{V}\int_{A_{li}} \mathbf{n}_{li}\rho_l D_l dA \tag{63}$$

which becomes

$$-\frac{1}{V}\int_{A_{ls}} \mathbf{n}_{ls}\rho_l D_l dA = -\frac{1}{V}\int_{A_{li}} \mathbf{n}_{li} \cdot (\rho_l D_l(\nabla \mathbf{b}_l + s_l \mathbf{I}))dA \tag{64}$$

because of Eq. (58). Then, Eq. (61) may be recast into a more convenient form

$$\rho_l \mathbf{u}_l = \frac{1}{V}\int_{A_{ls}} \mathbf{n}_{ls} \cdot \rho_l(\mathbf{v}_l - \mathbf{w}_{sl})\mathbf{b}_l dA - \frac{1}{V}\int_{A_{ls}} \mathbf{n}_{ls} \cdot \rho_l D_l(\nabla \mathbf{b}_l + s_l \mathbf{I})dA - \frac{1}{V}\int_{A_{li}} \mathbf{n}_{li} \cdot \rho_l D_l(\nabla \mathbf{b}_l + s_l \mathbf{I})dA \tag{65}$$

which can be used to check the consistency relationship

$$\rho_l \frac{\partial \langle \mathbf{b}_l \rangle}{\partial t} \equiv 0 \tag{66}$$

In the above closure problems, the zero average conditions on s_l and $\mathbf{b}_l$ make sure that the deviation of the mass fraction is zero. The initial conditions are essential to make the problem complete, but they are still not specified. Since they are dependent on the specific problem to be solved in the transient case, while in the fllowing we only consider the steady-state behavior, this discussion is left open here.

We can notice that the two problems are still coupled by the terms involving s_l. It is believed essential to keep such a coupling to better estimate the flux exchange. Up to this point, the interface velocity has been kept in the equations, which makes the mass balance equations consistent. However, if we want to solve these problems, the interface position, interface velocity, and the macro-scale mass fraction have to be known properties. Assuming that mass fluxes near the interface are triggered by the diffusive flux, which is acceptable in most dissolution problems, approximate closure problems may be developed as done in Ref. [29].

3.3 Macro-scale equations and effective parameters

Substituting the deviation of the mass fraction Eq. (47) in the averaged equation Eq. (33), we get

$$\frac{\partial \varepsilon_l \rho_l \Omega_l}{\partial t} + \nabla \cdot (\varepsilon_l \rho_l \Omega_l \mathbf{U}_l) + \nabla \cdot \left(\rho_l \langle \tilde{\mathbf{v}}_l s_l \rangle (\Omega_l - \omega_{eq}) + \rho_l \langle \tilde{\mathbf{v}}_l \mathbf{b}_l \rangle \cdot \nabla \Omega_l\right)$$
$$= \nabla \cdot (\varepsilon_l \rho_l D_l \nabla \Omega_l) + \nabla \cdot \left(\frac{1}{V}\int_{A_{ls}} \mathbf{n}_{ls} \rho_l D_l \left(\mathbf{b}_l \cdot \nabla \Omega_l + s_l (\Omega_l - \omega_{eq})\right) dA\right) \quad (67)$$
$$+ \nabla \cdot \left(\frac{1}{V}\int_{A_{li}} \mathbf{n}_{li} \rho_l D_l \left(\mathbf{b}_l \cdot \nabla \Omega_l + s_l (\Omega_l - \omega_{eq})\right) dA\right) - K_c$$

with K_c defined by Eq. (34).

It is convenient to write Eq. (67) under the form of a generalized dispersion equation such as

$$\frac{\partial \varepsilon_l \rho_l \Omega_l}{\partial t} + \nabla \cdot \left(\varepsilon_l \rho_l \Omega_l \mathbf{U}_l + \varepsilon_l \rho_l (\Omega_l - \omega_{eq}) \mathbf{U}_l^*\right) = \nabla \cdot \left(\varepsilon_l \rho_l \mathbf{D}_l^* \cdot \nabla \Omega_l\right) - K_c \quad (68)$$

with the dispersion tensor given by

$$\mathbf{D}_l^* = D_l\left(\mathbf{I} + \frac{1}{V_l}\int_{A_{ls}} \mathbf{n}_{ls} \mathbf{b}_l dA + \frac{1}{V_l}\int_{A_{li}} \mathbf{n}_{li} \mathbf{b}_l dA\right) - \langle \tilde{\mathbf{v}}_l \mathbf{b}_l \rangle^l \quad (69)$$

and the non-traditional effective velocity given by

$$\mathbf{U}_l^* = \langle \tilde{\mathbf{v}}_l s_l \rangle^l - \frac{1}{V_l}\int_{A_{ls}} \mathbf{n}_{ls} D_l s_l dA - \frac{1}{V_l}\int_{A_{li}} \mathbf{n}_{li} D_l s_l dA \quad (70)$$

The mass exchange term can be calculated as (neglecting higher-order derivatives than $\nabla \Omega_l$)

$$K_c = \Omega_{ls} K_s + (\Omega_l - \omega_{eq}) \frac{1}{V}\int_{A_{ls}} \mathbf{n}_{ls} \cdot \left(\rho_l s_l (\mathbf{v}_l - \mathbf{w}_{sl}) - \rho_l D_l \nabla s_l\right) dA$$
$$+ \left(\frac{1}{V}\int_{A_{ls}} \mathbf{n}_{ls} \cdot \left(\rho_l (\mathbf{v}_l - \mathbf{w}_{sl}) \mathbf{b}_l - \rho_l D_l (\nabla \mathbf{b}_l + s_l \mathbf{I})\right) dA\right) \cdot \nabla \Omega_l \quad (71)$$
$$- \rho_l D_l \left(\frac{1}{V}\int_{A_{ls}} \mathbf{n}_{ls} dA\right) \cdot \nabla \Omega_l$$

This equation is important because it illustrates the structure of the mass exchange term, no mater the boundary condition is reactive or thermodynamic equilibrium. Particularly, it indicates that the mass exchange term is dependent on both the the averaged concentration and its gradient. Moreover, if relate this term to the reaction rate expression of Eq. (34), we obtain the expression below

$$K_c = -a_{vl} M_c k_s \left\langle \left(1 - \frac{\Omega_l + s_l (\Omega_l - \omega_{eq}) + \mathbf{b}_l \cdot \nabla \Omega_l}{\omega_{eq}}\right)^n \right\rangle_{ls} \quad (72)$$

Like Eq. (45), the linearized version can be written as

$$K_c = -a_{vl} \frac{M_c k_s}{\omega_{eq}} \left(1 - \frac{\Omega_l}{\omega_{eq}}\right)^{n-1} (\omega_{eq} - \Omega_l - n\Omega_{ls}) \quad (73)$$

or

$$K_c = -a_{vl}M_c\left(1-\frac{\Omega_l}{\omega_{eq}}\right)^n k_{s,eff} + a_{vl}M_c\left(1-\frac{\Omega_l}{\omega_{eq}}\right)^{n-1}\mathbf{h}_l^* \cdot \nabla\Omega_l \tag{74}$$

where the effective reaction rate coefficient is given by

$$k_{s,eff} = k_s\left(1+n\langle s_l\rangle_{ls}\right) \tag{75}$$

in a similar form as in Ref. [39] for the first-order reaction case. The additional gradient term coefficient is

$$\mathbf{h}_l^* = n\frac{k_s}{\omega_{eq}}\langle \mathbf{b}_l\rangle_{ls} \tag{76}$$

One must remember that the macro-scale problem also involves the following averaged equations

$$\frac{\partial \varepsilon_l \rho_l}{\partial t} + \nabla\cdot(\rho_l \mathbf{V}_l) = -K_s, \quad \frac{\partial \varepsilon_s \rho_s}{\partial t} = K_s \tag{77}$$

The resulting macro-scale model is a generalization to nonlinear reaction rates of non-equilibrium models previously published. Such models are useful in many circumstances, for instance as diffuse interface models for modeling the dissolution of cavities [11, 40] or for studying instability of dissolution fronts [5].

4. Numerical examples

In this section, we validate the Darcy-scale model developed in the above section using the dissolution of a pillar located in a Hele-Shaw structure as an example and the computations were performed with the finite element software COMSOL Multiphysics®. Only half of the symmetric geometries were used in the simulations for the sake of simplicity, as presented in **Figure 3**. In the direct numerical simulations using the pore-scale model, the geometry is three-dimensional as shown in **Figure 3(a)**, with Poiseuille flow (average velocity of U_0) and zero mass fraction at the inlet (left surface), thermodynamic equilibrium mass fraction and no-slip boundary condition at the pillar surface (curved), zero mass flux and no-slip boundary conditions at the top, bottom and back surfaces, convective flux, and zero pressure at the outlet (right surface). After vertical averaging, the 3D geometry became 2D as presented in **Figure 3(b)**. Moreover, the whole medium was considered as a continuum in Darcy-scale model, different from at pore-scale that the solid and liquid spaces were considered separately. The pore-scale boundary condition at the solid/fluid interface was incorporated into the effective parameters when upscaled into Darcy-scale. Therefore, boundary conditions were only prescribed for other borders and they were the same as done in pore-scale modeling. In some previous studies [1, 12, 41], it was demonstrated that Darcy-Brinkmann equation

$$-\left(\nabla P_l - \rho_l \mathbf{g}\right) + \mu_l^* \nabla^2 \mathbf{V}_l - \mu_l \mathbf{K}_l^{-1}\cdot \mathbf{V}_l = 0 \tag{78}$$

is a better choice of the macro-scale momentum equation than Darcy's law, because it includes both a viscous and a Darcy term. When the permeability tends to infinity the equation simplifies to Stokes equation

$$-(\nabla P_l - \rho_l \mathbf{g}) + \mu_l^* \nabla^2 \mathbf{V}_l = 0 \tag{79}$$

and when the permeability is small enough Darcy's equation is recovered

$$\mathbf{V}_l = -\frac{\mathbf{K}_l}{\mu_l} \cdot (\nabla P_l - \rho_l \mathbf{g}) \tag{80}$$

For the computations of these two systems, without losing generality, we use pure gypsum dissolution in water as an example while it could also be other soluble materials, such as halite and carbonates in geological structures. The parameters used are presented in **Table 1** and the thermodynamic equilibrium mass fraction of the followed Ca^{2+} is estimated as

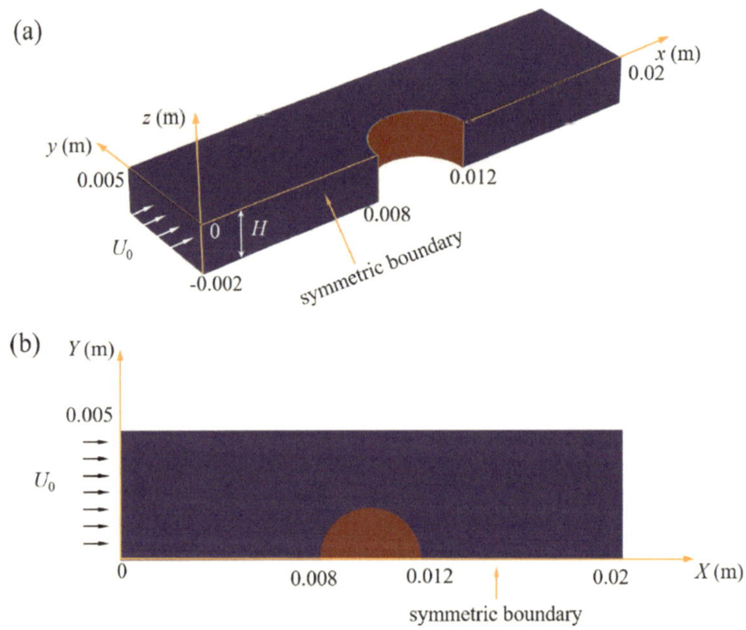

Figure 3.
The 3D geometry used for the DNS (a) and the 2D geometry used for the macro-scale model (b).

Parameter	Value
ρ_l	1000 kg m^{-3}
ρ_s	2310 kg m^{-3}
μ_l	10^{-3} Pa s
D_l	10^{-9} m^2 s^{-1}
L_r	2 mm

Table 1.
Simulation parameters.

$$\omega_{eq} = M_{Ca}\left(1.32 \times 10^{-2} + 1.31 \times 10^{-4}T - 1.47 \times 10^{-6}T^2\right) \tag{81}$$

where M_{Ca} is the molar weight of Ca and T is the common temperature in °C.

To describe the conditions used for the simulations, we introduce Darcy-scale Reynolds number Re_M and Péclet number Pe_M, i.e.,

$$\mathrm{Re}_M = \frac{\rho_r U_0 L_r}{\mu_l}; \mathrm{Pe}_M = \frac{U_0 L_r}{D_l} \tag{82}$$

where L_r is a characteristic length such as the aperture of the Hele-Shaw structure. These two dimensionless parameters were changed by varying the inlet velocity U_0 from 10^{-6}m s^{-1} to 5×10^{-2} m s^{-1}.

Three cross sections, namely the bottom, the middle and the top of the pillar obtained from DNS, are compared with Darcy-scale computations, after dissolution for 10^5 s. As shown in **Figure 4(a)–(c)**, the pillar first maintains circular cross sections and dissolves uniformly as observed at small Pe_M and Re_M. Then, the increases of Pe_M and Re_M lead to the formation of a small cusp in the downstream of the pillar (see **Figure 4(d)**), where dissolution is slower than the upstream of the pillar because the liquid is much more saturated with dissolved minerals, which agrees well with the experimental observations reported [12, 42, 43]. However, when Pe_M reaches 10^5 (and Re_M reaches 100), the dissolution in the downstream of the pillar is flattened, as also observed in Ref. [44]. This is because that in the case of large Re_M, the back side of the pillar is within the region of steady-state inertia vortices that were observed to appear for a large enough Reynolds number, where the solute is well mixed rapidly. Comparing the streamlines as shown in **Figure 5**, we see that, unlike the attached flow in the case of small Pe_M and Re_M, there appear indeed steady vortices at the backside of the pillar when Pe_M and Re_M are relatively large (indicated by the arrow in **Figure 5 (b)**).

Regarding the accuracy of the Darcy-scale model, one can observe that when Pe_M and Re_M are small, i.e., the mass transport process is dominated by diffusion and

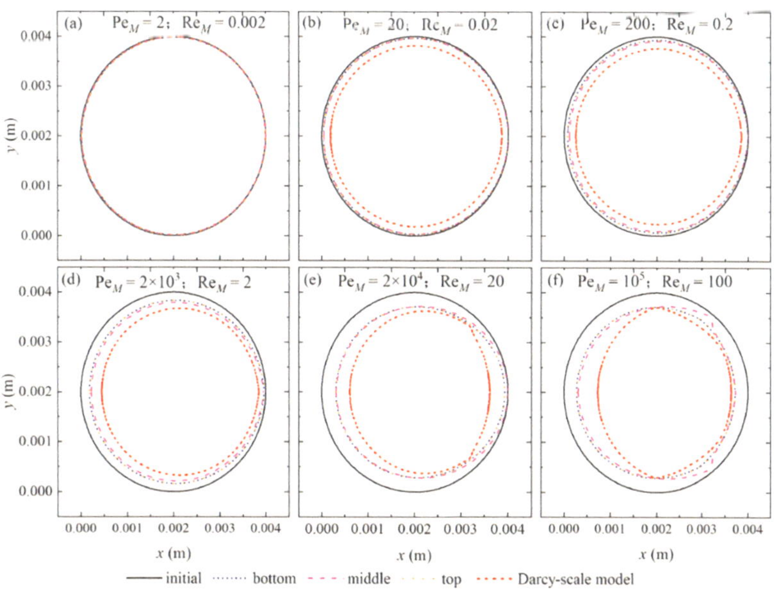

Figure 4.
Geometry comparison between Darcy-scale model results and cross sections at three levels of the gypsum pillar after dissolution of 10^5 *s for different* Pe_M *and* Re_M.

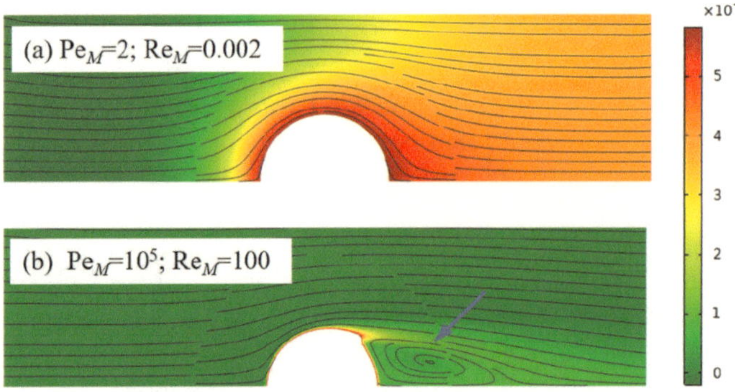

Figure 5.
Surface plotting of the mass fraction of Ca^{2+} *and corresponding streamlines.*

inertia and 3D effects are negligible, there is little difference between the size and circular shape of the three cross sections, which are remarkably well described by the Darcy model (see **Figure 4(a)–(d)**). However, when $Pe_M \geq 2 \times 10^4$, even though the variation in the vertical direction is not profound, the Darcy-scale model failed to capture either the position or the shape of the dissolution interface and this discrepancy is increased dramatically from the condition with $Pe_M = 2 \times 10^4$ ($Re_M = 20$) to that with $Pe_M = 10^5$ ($Re_M = 100$). This can be attributed to the fact that such conditions go beyond the assumption of small Reynolds number to develop the Darcy-scale model. In addition, as discussed in our previous work [1], to make sure the macro-scale model correctly reproduce the dissolution flux, the diffuse interface must not significantly affect the velocity and the mass fraction boundary layers, which means that the thickness of the diffuse interface (denoted as δ_D), which can be defined by the zone where the porosity varies between the two limits, must be much smaller than the thicknesses of the velocity boundary layer (δ_v) and the mass fraction boundary layer (δ_ω). From the results presented in **Figure 6**, one sees that this is indeed the case for diffusion dominated case, where we have $\delta_D \approx 0.0001$ m, $\delta_v \approx 0.0015$ m and $\delta_\omega \approx 0.003$ m, giving $\delta_D \ll \delta_v, \delta_\omega$. However, in the convection dominated case (see **Figure 6**), such

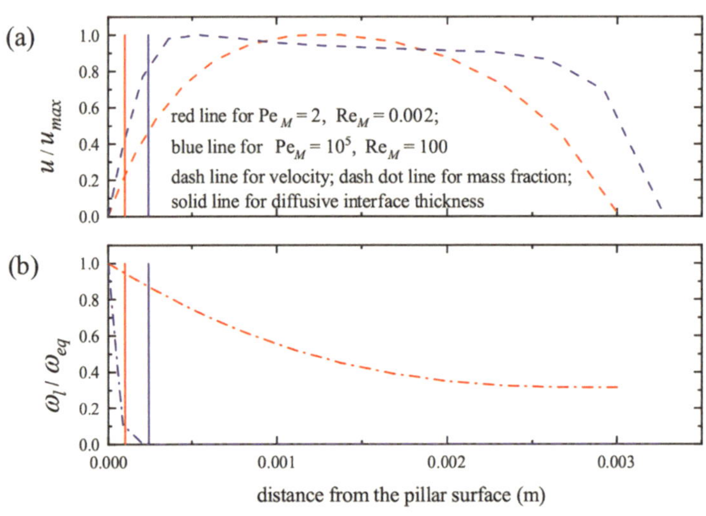

Figure 6.
Comparison of the velocity and mass fraction boundary layer thickness with that of the diffusive interface.

a hypothesis breaks down, where the three lengths are comparable, i.e., $\delta_D \approx \delta_\omega \approx 0.0002$ m and $\delta_v \approx 0.0005$ m. The above results demonstrate that there is an upper limit of the applicability range of the Dary-scale model in terms of Péclet and Reynolds numbers, i.e., $\mathcal{O}(10^3)$ and $\mathcal{O}(1)$, respectively, for the case considered here.

5. Conclusions

In this chapter, the upscaling of a mass transport problem involving a nonlinear heterogeneous reaction typical of dissolution problems has been carried out using a first-order Taylor's expansion for the reaction rate when developing the equations for the concentration deviation. A full model including all couplings and the interface velocity has been obtained, and the closure problems providing the effective properties have been presented. To validate the Darcy-scale model, we have investigated the case of the dissolution of a cylindrical pillar located in a Hele-Shaw structure. In principle, the resulting pillar shape during the dissolution process becomes three dimensional. However, the shape remains nearly cylindrical for low Péclet and Reynolds numbers. It was found that a Darcy-Brinkman formulation with a no-slip condition on the pillar surface and the same dispersion equation managed to reproduce the pillar surface evolution up to relatively large Péclet and Reynolds numbers.

Additional information

Parts of this chapter were previously published in the doctoral thesis by the same author: Jianwei Guo. Numerical modelling of the dissolution of karstic cavities. 2015. Institut National Polytechnique de Toulouse. Available from: http://ethesis.inp-toulouse.fr/archive/00003105/.

Thanks

The author expresses her thanks to the financial supports from the National Natural Science Foundation of China (Grant 12102371), the Natural Science Foundation of Sichuan Province, China (Grant 2022NSFSC1932) and the Fundamental Research Funds for the Central Universities (Grant 2682022KJ049).

Author details

Jianwei Guo
Southwest Jiaotong University, China

*Address all correspondence to: jianweiguo@swjtu.edu.cn

References

[1] Guo J, Laouafa F, Quintard M. A theoretical and numerical framework for modeling gypsum cavity dissolution. International Journal for Numerical and Analytical Methods in Geomechanics. 2016;**40**:1662-1689

[2] Cooper A. Halite karst geohazards (natural and man-made) in the United Kingdom. Environmental Geology. 2002;**42**(5):505-512

[3] Hao Y, Smith M, Carroll S. Multiscale modeling of CO2-induced carbonate dissolution: From core to meter scale. International Journal of Greenhouse Gas Control. 2019;**88**:272-289

[4] Li Q, Lin Z, Cai WH, Chen C-Y, Meiburg E. Dissolution-driven convection of low solubility fluids in porous media. International Journal of Heat and Mass Transfer. 2023;**217**: 124624

[5] Golfier F, Zarcone C, Bazin B, Lenormand R, Lasseux D, Quintard M. On the ability of a Darcy-scale model to capture wormhole formation during the dissolution of a porous medium. Journal of Fluid Mechanics. 2002;**457**:213-254

[6] Kalia N, Balakotaiah V. Effect of medium heterogeneities on reactive dissolution of carbonates. Chemical Engineering Science. 2009;**64**(2): 376-390

[7] Liu P, Yan X, Yao J, Sun S. Modeling and analysis of the acidizing process in carbonate rocks using a two-phase thermal-hydrologic-chemical coupled model. Chemical Engineering Science. 2019;**207**:215-234

[8] Panga MKR, Ziauddin M, Balakotaiah V. Two-scale continuum model for simulation of wormholes in carbonate acidization. AICHE Journal. 2005;**51**(12): 3231-3248

[9] Li X, Yang X. Effects of physicochemical properties and structural heterogeneity on mineral precipitation and dissolution in saturated porous media. Applied Geochemistry. 2022;**146**:105474

[10] Békri S, Thovert JF, Adler PM. Dissolution of porous media. Chemical Engineering Science. 1995;**50**:2765-2791

[11] Luo H, Quintard M, Debenest G, Laouafa F. Properties of a diffuse interface model based on a porous medium theory for solid-liquid dissolution problems. Computational Geosciences. 2012;**16**(4):913-932

[12] Soulaine C, Roman S, Kovscek A, Tchelepi HA. Mineral dissolution and wormholing from a pore-scale perspective. Journal of Fluid Mechanics. 2017;**827**:457-483

[13] Soulaine C, Roman S, Kovscek A, Tchelepi HA. Pore-scale modelling of multiphase reactive flow: Application to mineral dissolution with production of CO2. Journal of Fluid Mechanics. 2018; **855**:616-645

[14] Starchenko V, Marra CJ, Ladd AJC. Three-dimensional simulations of fracture dissolution. Journal of Geophysical Research Solid Earth. 2016; **121**(9):6421-6444

[15] Olsson E, Kreiss G. A conservative level set method for two phase flow. Journal of Computational Physics. 2005; **210**(1):225-246

[16] Olsson E, Kreiss G, Zahedi S. A conservative level set method for two

phase flow II. Journal of Computational Physics. 2007;**225**:785-807

[17] Li H, Wang F, Wang Y, Yuan Y, Feng G, Tian H, et al. Phase-field modeling of coupled reactive transport and pore structure evolution due to mineral dissolution in porous media. Journal of Hydrology. 2023;**619**:129363

[18] Békri S, Renard S, Delprat-Jannaud F. Pore to core scale simulation of the mass transfer with mineral reaction in porous media. Oil & Gas Science and Technology–Revue d'IFP Energies nouvelles. 2015;**70**:681-693

[19] Varloteaux C, Békri S, Adler PM. Pore network modelling to determine the transport properties in presence of a reactive fluid: From pore to reservoir scale. Advances in Water Resources. 2013;**53**:87-100

[20] Vignoles GL, Aspa Y, Quintard M. Modelling of carbon-carbon composite ablation in rocket nozzles. Composites Science and Technology. 2010;**70**(9): 1303-1311

[21] Taylor G. Dispersion of soluble matter in solvent flowing slowly through a tube. Proceedings of the Royal Society of London. Series A. Mathematical and Physical Sciences. 1953;**219**(1137): 186-203

[22] Taylor G. The dispersion of matter in turbulent flow through a pipe. Proceedings of the Royal Society of London. Series A. Mathematical and Physical Sciences. 1954;**223**(1155): 446-468

[23] Aris R. On the dispersion of a solute in a fluid flowing through a tube. Proceedings of the Royal Society of London. Series A. Mathematical and Physical Sciences. 1956;**235**(1200):67-77

[24] Brenner H, Stewartson K. Dispersion resulting from flow through spatially periodic porous media. Philosophical Transactions of the Royal Society of London. Series A, Mathematical and Physical Sciences. 1980;**297**(1430): 81-133

[25] Eidsath A, Carbonell RG, Whitaker S, Herrmann LR. Dispersion in pulsed systems-III: Comparison between theory and experiments for packed beds. Chemical Engineering Science. 1983;**38** (11):1803-1816

[26] Mei CC. Method of homogenization applied to dispersion in porous media. Transport in Porous Media. 1992;**9**(3): 261-274

[27] Bousquet-Melou P, Neculae A, Goyeau B, Quintard M. Averaged solute transport during solidification of a binary mixture: Active dispersion in dendritic structures. Metallurgical & Materials Transactions B. 2002;**33**(3): 365-376

[28] Coutelieris FA, Kainourgiakis ME, Stubos AK, Kikkinides ES, Yortsos YC. Multiphase mass transport with partitioning and inter-phase transport in porous media. Chemical Engineering Science. 2006;**61**(14):4650-4661

[29] Guo J, Quintard M, Laouafa F. Dispersion in porous media with heterogeneous nonlinear reactions. Transport in Porous Media. 2015;**109**(3): 541-570

[30] Quintard M, Whitaker S. Convection, dispersion, and interfacial transport of contaminants: Homogeneous porous media. Advances in Water Resources. 1994;**17**:221-239

[31] Whitaker S. The Method of Volume Averaging. Dordrecht, The Netherlands: Kluwer Academic Publishers; 1999

[32] Cushman J, Ginn TR. Nonlocal dispersion in media with continuously evolving scales of heterogeneity. Transport in Porous Media. 1993;**13**: 123-138

[33] Dagan G. Flow and Transport in Porous Formations. Berlin-Heidelberg: Springer; 1989

[34] Brenner H. Dispersion resulting from flow through spatially periodic porous media. Philosophical Transactions of the Royal Society of London. 1980;**297**:81-133

[35] Shapiro M, Brenner H. Dispersion of a chemically reactive solute in a spatially periodic model of a porous medium. Chemical Engineering Science. 1988;**43**: 551-571

[36] Bensoussan A, Lions JL, Papanicolau G. Asymptotic Analysis for Periodic Structures. Amsterdam: North-Holland Publishing Company; 1978

[37] Soulaine C, Debenest G, Quintard M. Upscaling multi-component two-phase flow in porous media with partitioning coefficient. Chemical Engineering Science. 2011;**66**:6180-6192

[38] Valdés-Parada FJ, Aguilar-Madera CG, J. Álvarez-Ramírez on diffusion, dispersion and reaction in porous media. Chemical Engineering Science. 2011;**66** (10):2177-2190

[39] Wood BD, Radakovich K, Golfier F. Effective reaction at a fluid-solid interface: Applications to biotransformation in porous media. Advances in Water Resources. 2007;**30**: 1630-1647

[40] Luo H, Laouafa F, Guo J, Quintard M. Numerical modeling of three-phase dissolution of underground cavities using a diffuse interface model. International Journal for Numerical and Analytical Methods in Geomechanics. 2014;**38**:1600-1616

[41] Guo J, Laouafa F, Quintard M. On 2d approximations for dissolution problems in hele-Shaw cells. Science China-Physics, Mechanics and Astronomy. 2023;**66**:234711

[42] Dutka F, Starchenko V, Osselin F, Magni S, Szymczak P, Ladd AJC. Time-dependent shapes of a dissolving mineral grain: Comparisons of simulations with microfluidic experiments. Chemical Geology. 2020;**540**:119459

[43] Tan Q, Kang Y, You L, Peng H, Chen Q. Pore-scale investigation on mineral dissolution and evolution in hydrological properties of complex porous media with binary minerals. Chemical Geology. 2023;**616**:121247

[44] Huang JM, Moore MNJ, Ristroph L. Shape dynamics and scaling laws for a body dissolving in fluid flow. Journal of Fluid Mechanics. 2015;**765**:R3

Chapter 9

Enhancement of Mass Transfer and Coke Resistance in DRM through Hierarchical Porous Catalysts

Yixiong Lin, Chen Yang and Ting Qiu

Abstract

Dry reforming of methane (DRM) is one of the feasible strategies for carbon capture and utilization. However, DRM has a high tendency toward coking, which is restricted to industrial applications. The primary cause of coking in DRM is the limitation of mass transfer inside porous catalysts. To overcome this limitation, optimizing the pore structure of the porous catalyst becomes crucial. Hierarchical pore structure has received considerable attention in recent years due to its superior mass transfer performance. Therefore, this chapter focuses on the structure-performance relationship of hierarchical porous catalysts in DRM. Specifically, two types of porous catalysts, namely porous pellet and open-cell foam, are examined. The impacts of various hierarchical pore structure parameters on the catalytic activity and coke resistance are investigated. The findings offer a theoretical foundation and technical guidance for the design of porous catalysts with hierarchical pore structures.

Keywords: dry reforming of methane, porous pellet, open-cell foam, hierarchical pore structures, mass transfer, coke resistance

1. Introduction

Dry reforming of methane (DRM) offers a promising solution by converting two primary greenhouse gases, carbon dioxide (CO_2) and methane (CH_4), into value-added synthesis gas (H_2 and CO), which can be applied to synthesize oxygenated chemicals [1] and hydrocarbons *via* Fischer-Tropsch synthesis [2]. Due to their remarkable cost-effectiveness and excellent reaction performance, catalysts based on nickel (Ni) have gained widespread utilization in DRM. However, the endothermic nature of the DRM process requires high temperature (>873.15 K) [3], leading to a high tendency toward carbon deposition on Ni-based catalysts, resulting in rapid deactivation and hindering industrial applications [4, 5].

To promote the industrial application of DRM, two effective methods have been pursued: active site design and catalyst pore structure optimization. In recent years, an increasing number of researchers have devoted their efforts to catalyst active site design, leading to notable enhancements in the activity and coke resistance of Ni-based catalysts. Various approaches, such as utilizing different support materials [6], incorporating promoters [7], and optimizing preparation and activation methods [8], have been employed to reinforce the coke resistance of Ni-based catalysts. These studies have made significant progress in improving the catalytic activity of DRM catalysts. Concurrently, the optimization of catalyst pore structure has emerged as a crucial strategy for mitigating catalyst deactivation. Notably, Rao et al. [9] have demonstrated that pore structure optimization can nearly double the lifespan time of catalyst and enhance coke resistance against deactivation. However, it is noteworthy that only limited research has been focused specifically on catalyst pore structure optimization.

In pursuit of enhancing the catalytic activity and resistance to coke formation, the catalyst pore structure has been studied in this work. Porous pellets are extensively employed in the DRM process. However, Baiker et al. [10] highlighted that the catalyst performance of porous pellets in many industrial processes is restricted by intraparticle diffusion despite their high catalytic activity. To overcome this limitation, the concept of hierarchical pore structures has garnered significant attention in recent years due to their superior mass transfer properties [11–13]. Notably, Lakiss et al. [14] observed that porous pellets with hierarchical pore structures exhibit shorter diffusional paths. As a result, numerous researchers have made efforts to prepare porous pellets with bimodal hierarchical pore structures, involving combinations of micro-mesopores [15], micro-macropores [16], and macro-mesopores [17, 18]. The hierarchical pore structure of porous pellets also demonstrates remarkable coke resistance. Despite these advancements, however, there remains a lack of comprehensive studies associated with the effects of porous pellets with hierarchical pore structures on both catalyst performance and coke resistance for DRM.

Additionally, structured catalysts present another category of porous catalyst worth considering for DRM. In contrast to unstructured catalysts like porous pellets, structured catalysts offer advantages by minimizing regions of restricted mass transfer, thereby mitigating catalyst deactivation [19, 20]. Among the structured catalysts, open-cell foam stands out for its widespread use in heterogeneous reactions, owing to its high porosity, and permeability, as well as excellent heat and mass transfer performance. Notably, Richardson et al. [21] conducted experimental studies confirming that foam catalysts exhibited a higher effectiveness factor than commercially available porous pellets in DRM. Furthermore, investigations have revealed that foam catalysts maintain comparable catalyst performance and selectivity to powdered pellet catalysts. Consequently, many researchers have explored the use of open-cell foam as a catalyst in DRM [22–24]. Based on the superior mass transfer performance of the hierarchical pore structure, however, a critical knowledge gap remains as the impact of open-cell foam with hierarchical pore structure on coke resistance during the DRM process has not been thoroughly explored yet. Understanding this aspect can elucidate the potential advantages and limitations of employing such structured catalysts, thereby contributing to the advancement of open-cell foam structure catalysts.

This work aims to investigate the relationship between porous catalysts with hierarchical pore structure and catalyst performance during DRM. To achieve this, a numerical model is employed to analyze the mass transfer and DRM reaction in porous pellet and open-cell foam. Moreover, two artificial numerical algorithms are developed specifically for constructing hierarchical pore structures for porous pellet and open-cell foam, respectively. Based on Benguerba et al. [25], five reactions associated with DRM shown in **Table 1** are used in the study, where R_3, R_4 and R_5 are related to coking. The impacts of various hierarchical pore structure parameters on the catalyst performance and coke resistance are investigated.

Index	Reactions	Reaction rates
R_1	$CO_2 + CH_4 \leftrightarrow 2CO + 2H_2$	$R_1 = \frac{k_1 K_{CO_2,1} K_{CH_4,1} P_{CH_4} P_{CO_2}}{\left(1+K_{CO_2,1} P_{CO_2} + K_{CH_4,1} P_{CH_4}\right)^2} \left(1 - \frac{\left(P_{CO} P_{H_2}\right)^2}{K_{P_1} P_{CH_4} P_{CO_2}}\right)$
R_2	$CO_2 + H_2 \leftrightarrow CO + H_2O$	$R_2 = \frac{k_2 K_{CO_2,2} K_{H_2,2} P_{H_2} P_{CO_2}}{\left(1+K_{CO_2,2} P_{CO_2} + K_{H_2,2} P_{H_2}\right)^2} \left(1 - \frac{\left(P_{CO} P_{H_2O}\right)^2}{K_{P_2} P_{CO_2} P_{H_2}}\right)$
R_3	$CH_4 \leftrightarrow C + 2H_2$	$R_3 = \frac{k_e K_{CH_4,3} \left(P_{CH_4} - \frac{P_{H_2}^2}{K_{P_3}}\right)}{\left(1+K_{CH_4,3} P_{CH_4} + \frac{P_{H_2}^{1.5}}{K_{H_2,3}}\right)^2}$
R_4	$C + H_2O \leftrightarrow CO + H_2$	$R_4 = \frac{\frac{k_4}{K_{H_2O}} \left(\frac{P_{H_2O}}{P_{H_2}} - \frac{P_{CO}}{K_{P_4}}\right)}{\left(1+K_{CH_4,4} P_{CH_4} + \frac{P_{H_2O}}{K_{H_2O,4} P_{H_2}} + \frac{P_{H_2}^{1.5}}{K_{H_2,4}}\right)^2}$
R_5	$C + CO_2 \leftrightarrow 2CO$	$R_5 = \frac{\frac{k_5}{K_{CO,5} K_{CO_2,5}} \left(\frac{P_{CO_2}}{P_{CO}} - \frac{P_{CO}}{K_{P_5}}\right)}{\left(1+K_{CO,5} P_{CO} + \frac{P_{CO_2}}{K_{CO_2,5} K_{CO,5} P_{CO}}\right)^2}$

Table 1.
Five reactions kinetics during DRM.

2. Porous pellet

The DRM process in a porous pellet is illustrated in **Figure 1**. In porous pellets, three kinds of intraparticle diffusion behaviors are considered in this work, which are self-diffusion, multicomponent diffusion, and Knudsen diffusion.

2.1 Reconstruction of porous pellet with hierarchical pore structure

In DRM, the utilization of porous pellets containing nickel-based catalysts with macro-mesoporous structures is prevalent [26, 27]. Several artificial algorithms to reconstruct macro-mesopore structures were developed by numerous researchers. Hussain et al. [28] introduced the random generation of macro-meso pores (RGMMP) algorithm, specifically designed for reconstructing porous building materials with interconnected macropores and mesopores in series. Another method proposed by Chen et al. [29] involves the random placement of circular solids to reconstruct

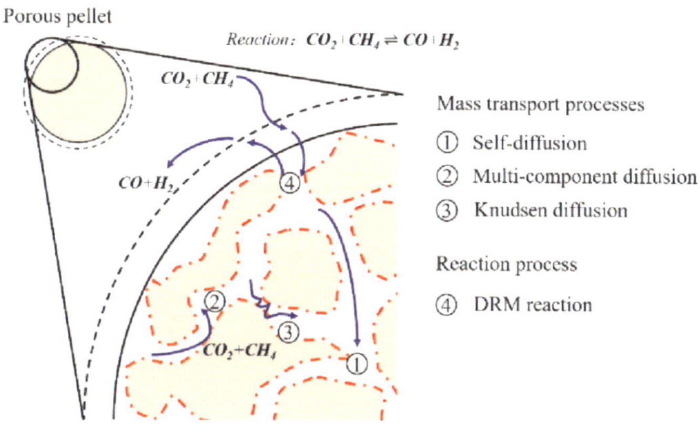

Figure 1.
Schematic diagram of the DRM process in a porous pellet.

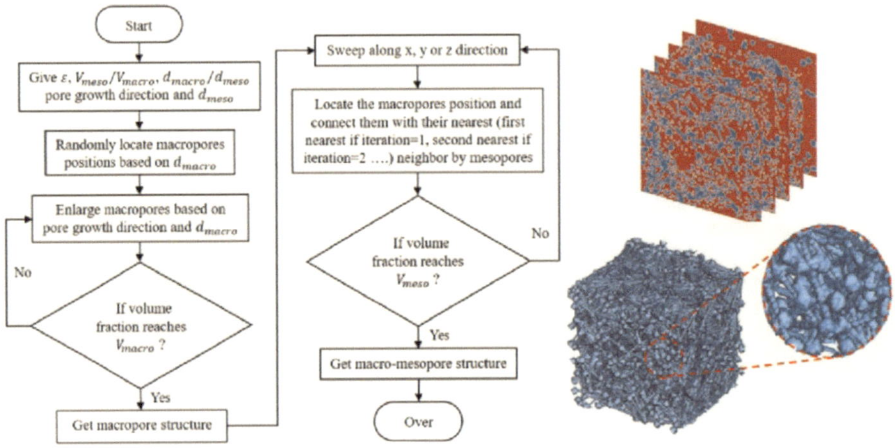

Figure 2.
Flowsheet of the modified RGMMP algorithm and reconstructed porous pellet with macro-mesopore structure [30].

hierarchical pore structures. While these algorithms offer promising capabilities, their paramcters are mathematical variables that may not directly construct hierarchical pore structures through experimental characterization data. To overcome this, Lin et al. [30] proposed a modified RGMMP algorithm, which can reconstruct hierarchical pore structures based on experimental characterization data. The modified RGMMP algorithm, depicted in **Figure 2**, is governed by five key variables, which are catalyst porosity (ε), the ratio of mesopore volume to macropore volume (V_{meso}/V_{macro}), the ratio of average macropore diameter to average mesopore diameter (d_{macro}/d_{meso}), and pore growth direction and mesopore diameter (d_{meso}).

2.2 Mass transfer in porous pellet with macro-mesopore structure

2.2.1 Effect of the ratio of mesopore volume to macropore volume

This section aims to explore intraparticle diffusivity while maintaining a constant total porosity and varying V_{meso}/V_{macro}. This study considered nine different ratios of V_{meso}/V_{macro} at values of 1/5, 1/4, 1/3, 1/2, 0.7, 0.9, 1.0, 1.2, and 1.4. **Figure 3** illustrates

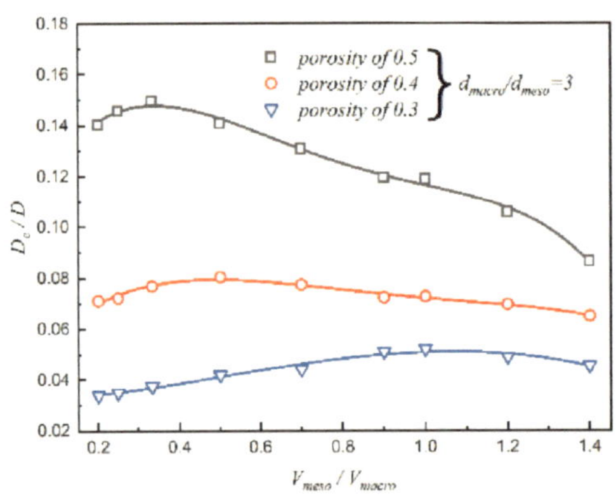

Figure 3.
Effect of V_{meso}/V_{macro} on intraparticle diffusivity [30].

the relationship between intraparticle diffusivity and the variation of V_{meso}/V_{macro}. Additionally, three catalyst porosities, namely ε = 0.3, 0.4, and 0.5, are examined. The results reveal an interesting trend that as the mesopore volume increases, the intraparticle diffusivity experiences a slight increase initially, but subsequently starts to decrease. This observation suggests the presence of an optimal value for V_{meso}/V_{macro}. The underlying reason for the initial rise in intraparticle diffusivity with increased mesopores is the enhanced interconnection of macropores facilitated by the presence of mesopores, which supports the diffusion process. However, as the mesopore volume continues to increase beyond the optimal point, the intraparticle diffusivity starts to decline. This can be attributed to the fact that while a certain amount of mesopores strengthens the interconnectivity of macropores, an excessive amount of mesopores reduces the number of available macropores, ultimately leading to the decline of intraparticle diffusivity. Furthermore, **Figure 3** shows that as the catalyst porosity decreases, the optimal value of V_{meso}/V_{macro} increases. This suggests that for lower porosity, a higher portion of mesopores is required to achieve higher intraparticle diffusivity.

2.2.2 Effect of the ratio of average macropore diameter to average mesopore diameter

To investigate the effect of d_{macro}/d_{meso}, we plotted the intraparticle diffusivity against the variation of d_{macro}/d_{meso} for V_{meso}/V_{macro} = 0.5, as shown in **Figure 4**. Eight ratios of d_{macro}/d_{meso} =1, 2, 3, 4, 5, 6, 8, and 10 were considered. From the observations in **Figure 4**, it is evident that the maximum intraparticle diffusivity is achieved when the value of d_{macro}/d_{meso} is approximately 4. This result aligns with the findings of Peng et al. [31], who emphasized that optimizing d_{macro}/d_{meso} leads to optimal mass transfer performance based on generalized Murray's law. When d_{macro}/d_{meso} is less than 4, increasing the macropore diameter enhances intraparticle diffusivity. This is because larger macropore diameters reduce diffusion resistance, as suggested by Wang et al. [32]. Nevertheless, when contemplating a constant quantity of macropore and mesopore volumes, increasing the average diameter of macropores results in larger voids between them. Consequently, to ensure adequate connectivity between neighboring macropores, an increased number of mesopores is required, which consequently causes a reduction in overall macropore interconnectivity. This, in turn, results in a decline in intraparticle diffusivity as the average macropore diameter

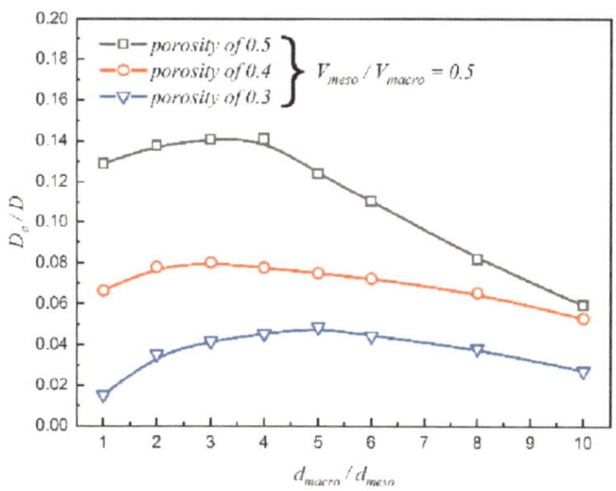

Figure 4.
Effect of V_{meso}/V_{macro} on intraparticle diffusivity [30].

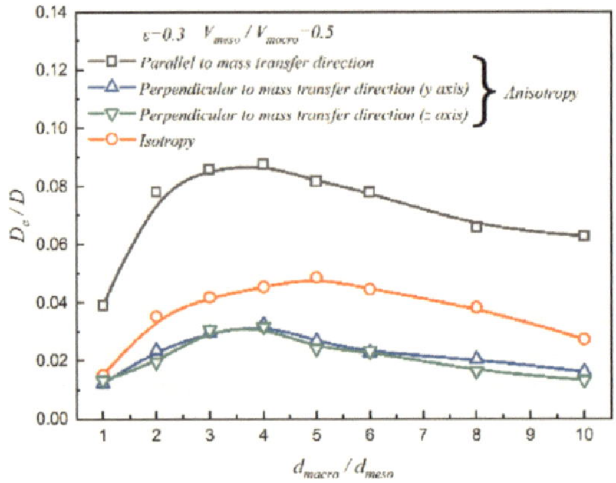

Figure 5.
Effect of pore growth direction on intraparticle diffusivity [30].

increases. These findings present valuable insights for optimizing the design of porous pellets through the manipulation of d_{macro}/d_{meso}.

2.2.3 Effect of macropore growth direction

In the modified RGMMP algorithm, mesopores are important for connecting macropores in porous pellets. This connection determines the growth direction in macro-mesopore structured pellets, which rely on the macropore structure. Four different types of macropore growth directions were defined in this study, which is isotropic, parallel to mass transfer direction, perpendicular to mass transfer direction in the *y*-axis, and perpendicular to mass transfer direction in the *z*-axis. **Figure 5** shows how intraparticle diffusivity changes with the ratio of average macropore diameter to average mesopore diameter (d_{macro}/d_{meso}), with ε =0.3 and V_{meso}/V_{macro} = 0.5. It can also be found that intraparticle diffusivities of parallel to the mass transfer direction and perpendicular to the mass transfer direction are the upper and lower bounds of the intraparticle diffusivity, respectively. Intraparticle diffusivity of perpendicular to the mass transfer direction in the *y*-axis is close to that of perpendicular to the mass transfer direction in the *z*-axis. For most cases, porous pellets exhibit the isotropic growth direction of macropore, whose intraparticle diffusivity is closer to the lower bound. It clearly indicates that the pore growth direction significantly affects intraparticle diffusivity under identical conditions.

2.3 DRM reaction in porous pellet with macro-mesopore structure

In DRM, this part aimed to explore the effects of three hierarchical pore parameters on coke resistance and catalyst performance. The investigation involved calculating several characterization parameters to assess the catalyst performance. These parameters were as follows:

$$R_{DRM} = \frac{\int_0^{N_T} r_{DRM} dN_T}{V} \tag{1}$$

$$R_{Coking} = \frac{\int_0^{N_T} r_{Coking} dN_T}{V} \tag{2}$$

$$N_{Active} = \frac{N_T}{V} \tag{3}$$

The larger the difference between R_{DRM} and R_{Coking}, the higher the catalyst performance. In order to investigate how the hierarchical structure of the catalyst affects its performance, we maintain the constant reaction operating conditions for DRM (T = 923.15 K, P = 1 bar, F_{CH_4}/F_{CO_2} = 1:1). These conditions align with Benguerba et al. [25].

2.3.1 Effect of catalyst porosity

The variations in catalyst porosities (ε = 0.3, 0.4, 0.5, 0.6, 0.7, 0.75, 0.8) are depicted in **Figure 6** and **Figure 7**, showing the reaction fluxes of R_{DRM} and R_{Coking}, as well as the corresponding catalyst performance. From **Figure 6**, it becomes evident that the optimal values of R_{DRM} and R_{Coking} occur when the porosity is approximately 0.6. This finding is attributed to the positive impact of adding porosity dominating at relatively low porosity levels (ε < 0.6), leading to an increase in R_{DRM} and R_{Coking}.

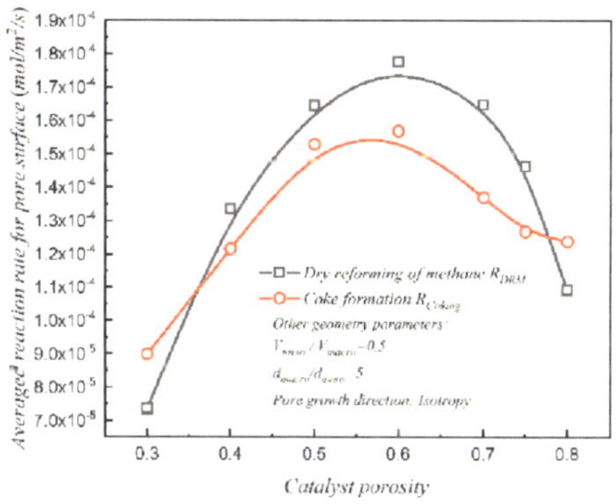

Figure 6.
Effect of ε on the reaction fluxes [33].

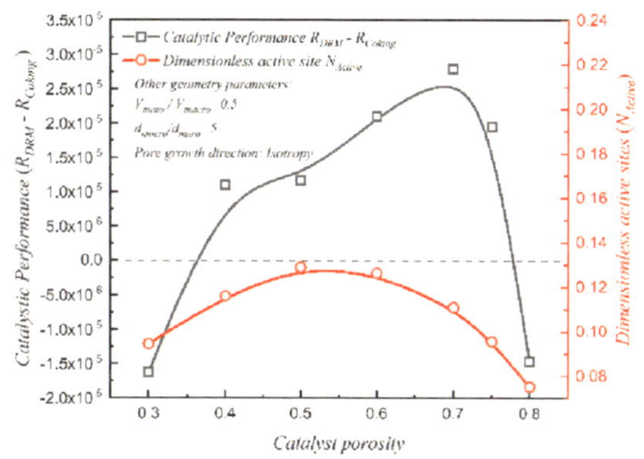

Figure 7.
Effect of ε on the catalyst performance [33].

However, when the catalyst porosity exceeds 0.6, both R_{DRM} and R_{Coking} decrease sharply. These observations suggest that the enhanced intraparticle mass transport cannot make up for the decrease in active sites. Moreover, it can be deduced from **Figure 7** that the active sites begin to decline once the porosity exceeds 0.6, indicating the competition between heterogeneous reaction and intraparticle diffusion. The trend observed in the variation of R_{Coking} aligns with R_{DRM}, resulting in the maximum R_{Coking} when R_{DRM} approaches its peak. As shown in **Figure 7**, it has been determined that optimal catalyst performance can be achieved by attaining a larger difference between R_{DRM} and R_{Coking}, which occurs when the catalyst porosity is close to 0.7. Therefore, for DRM, the optimal catalyst porosity is found to lie within the range of 0.6 to 0.7. Moreover, these findings are consistent with Liu et al. [26], who pointed out that the optimal porosity of catalysts with a macro-mesopore structure for DRM falls within the range of 0.61 to 0.64.

2.3.2 Effect of the ratio of mesopore volume to macropore volume

To gain a deeper understanding of the reaction-diffusion process in porous pellets with macro-mesopore structure, the pore volume becomes a crucial parameter while keeping the catalyst porosity constant. In this case, seven ratios were considered, namely V_{meso}/V_{macro} = 0.2, 0.3, 0.5, 0.8, 1.0, 1.25, and 1.5. As depicted in **Figure 8**, an interesting trend emerges as the mesopore volume increases. The average reaction fluxes of R_{DRM} and R_{Coking} gradually increase, reaching a peak when the ratio of mesopore volume to macropore volume (V_{meso}/V_{macro}) is equal to 0.5. However, as the mesopore volume continues to increase beyond this point, R_{DRM} and R_{Coking} drop dramatically. Furthermore, **Figure 9** confirms the observation that the optimal catalyst performance is achieved when the value of V_{meso}/V_{macro} approaches 0.5 as it maximizes the difference between R_{DRM} and R_{Coking}. Interestingly, it is evident from **Figure 9** that the active sites of the catalyst exhibit minimal variation with changes in V_{meso}/V_{macro}, indicating that this parameter has little effect on the active sites. Therefore, in this section, the catalyst performance is primarily limited by intraparticle diffusion. According to Lin et al. [30], the presence of an optimal value of V_{meso}/V_{macro} in the intraparticle diffusion process can be attributed to the increased mesopores, which enhance pore connectivity and reduce the number of dead pores.

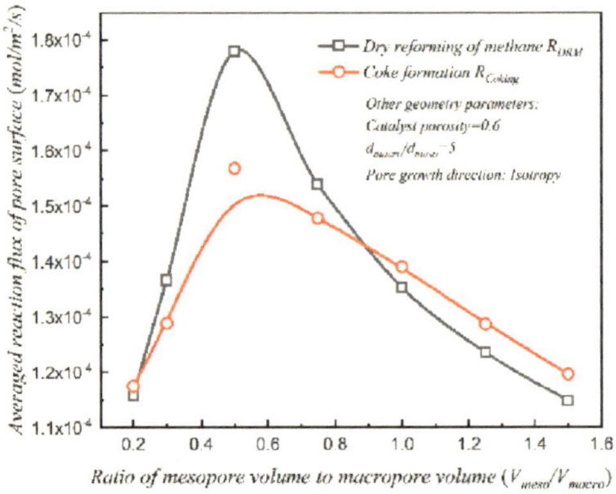

Figure 8.
Effect of V_{meso}/V_{macro} on the reaction fluxes [33].

DOI: http://dx.doi.org/10.5772/intechopen.1003689

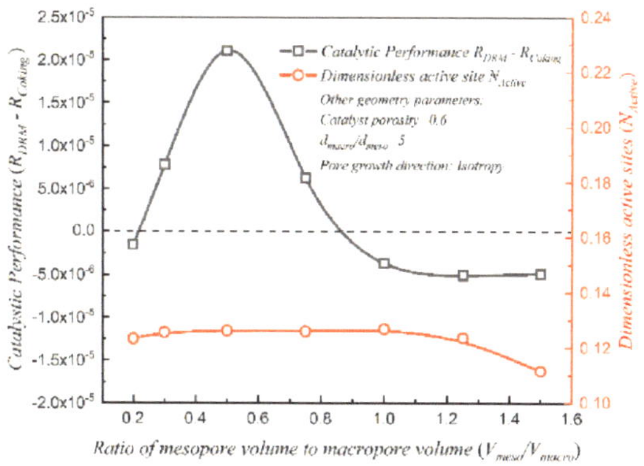

Figure 9.
Effect of V_{meso}/V_{macro} on the catalyst performance [33].

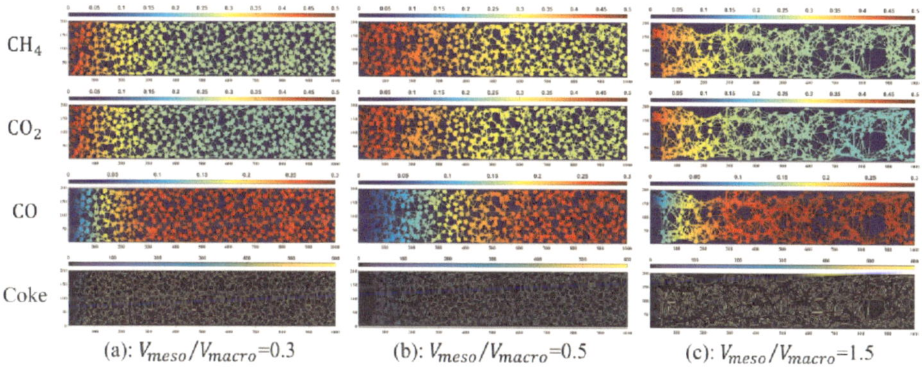

Figure 10.
Mole fraction distributions of CH_4, CO_2, CO, and amount distribution of coke (unit: Mol/m2/s) with three V_{meso}/V_{macro} [33]: (a) V_{meso}/V_{macro} =0.3; (b) V_{meso}/V_{macro} =0.5; (c) V_{meso}/V_{macro} =1.5.

Consequently, the catalyst performance is enhanced. However, if the mesopore volume continues to increase continuously, it leads to a reduction in the number of macropores, eventually weakening the catalyst performance. These findings indicate that an optimal balance of mesopore and macropore volumes is crucial for achieving the best catalyst performance. Too few mesopores limit the intraparticle diffusion process, while too many mesopores at the expense of macropores can reduce the overall catalytic activity. Thus, maintaining an appropriate ratio of mesopore volume to macropore volume is essential for optimizing the catalyst performance for macro-mesopore structure.

In **Figure 10**, mole fraction distributions of species and the amount distribution of coke are presented for different V_{meso}/V_{macro}. Evidently, when V_{meso}/V_{macro} = 0.5, the macro-mesopore structure exhibits optimal mass transport performance, resulting in reduced coke formation.

2.3.3 Effect of the ratio of average macropore diameter to average mesopore diameter

This section investigated the impact of the ratio between the average diameter of macropores and mesopores (d_{macro}/d_{meso}) on the catalyst performance. As depicted in

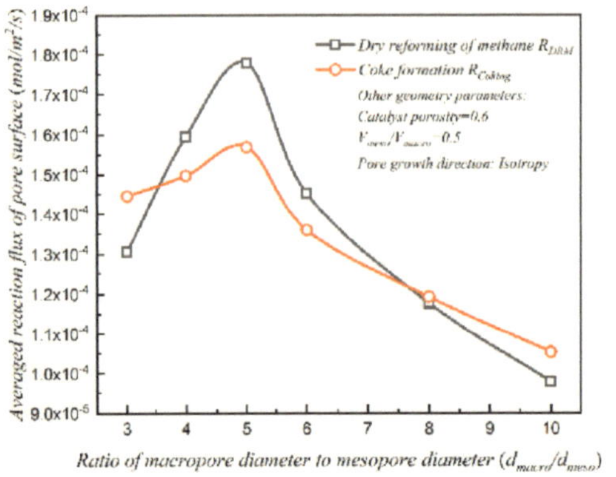

Figure 11.
Effect of d_{macro}/d_{meso} on reaction fluxes [33].

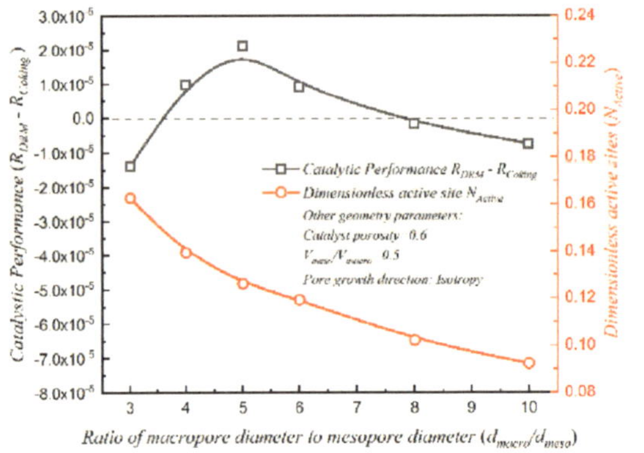

Figure 12.
Effect of d_{macro}/d_{meso} on the catalyst performance [33].

Figures 11 and **12**, six pore diameter ratios were considered, namely d_{macro}/d_{meso} = 3, 4, 5, 6, 8, and 10. Observing from **Figure 11**, it becomes evident that R_{DRM} reaches its optimal value as d_{macro}/d_{meso} approaches 5. Furthermore, R_{Coking} surpasses R_{DRM} when d_{macro}/d_{meso} is below 3 and above 8. Additionally, **Figure 12** illustrates that the maximum catalyst performance occurs at approximately d_{macro}/d_{meso} = 5. When d_{macro}/d_{meso} is below 5, increasing the macropore diameter effectively enhances the catalyst performance, as the positive impact of intraparticle diffusion outweighs the effect of diminishing active sites. However, as seen in **Figure 12**, a further increase in d_{macro}/d_{meso} from 5 to 10 results in a decline in macropore interconnectivity and active sites, ultimately leading to the degradation of the catalyst performance. The optimal value of d_{macro}/d_{meso} arises from the interplay between competitive heterogeneous reactions and intraparticle diffusion processes.

Figure 13 displays the mole fraction distributions of species, as well as the amount distribution of coke, corresponding to three different d_{macro}/d_{meso} =3, 5 and 10. In **Figure 13(b)**, the reactants CH_4 and CO_2, as well as the product CO, exhibit more extensive penetration lengths and lesser carbon deposition across the entire simulation domain compared to those shown in **Figure 13(a)** and **Figure 13(c)**.

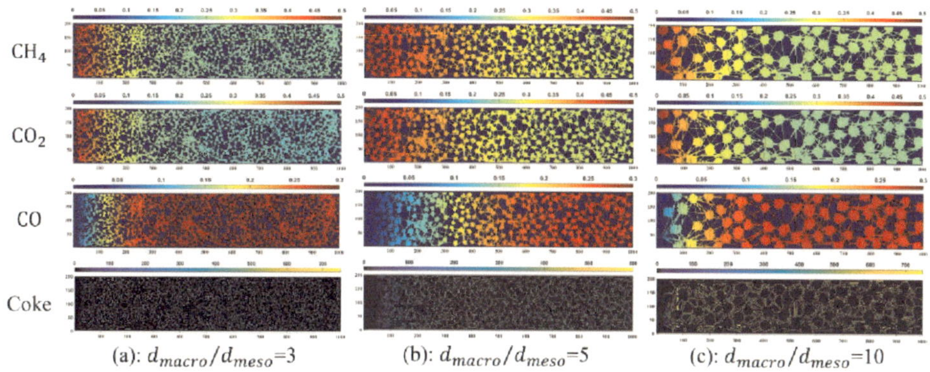

Figure 13.
Mole fraction distributions of CH_4, CO_2, CO, and amount distribution of carbon deposition (unit: Mol/m2/s) with three d_{macro}/d_{meso} [33]: (a) d_{macro}/d_{meso} =3; (b) d_{macro}/d_{meso} =5; (c) d_{macro}/d_{meso} =10.

3. Open-cell foam

This study examines the fluid flow, mass transport, and DRM reaction within open-cell foam, featuring hierarchical pore structure. As depicted in **Figure 14**, the convective-diffusion process only occurs in the pore area, while the DRM reaction and coke formation take place on the surface of the solid matrix.

3.1 Reconstruction of open-cell foam with hierarchical pore structure

Several idealized geometric models, including cubic models [34, 35], face-centered models, body-centered models [36], and Kelvin's tetrakaidecahedral model [37, 38], have been utilized to numerically investigate the pressure drop and heat transport characteristics. To construct hierarchical pore structure for open-cell foam, Lin et al. [39] proposed a novel numerical algorithm capable of constructing both uniform and hierarchical pore structures. Four parameters are used to control the algorithm, which are porosity (ε), fine pore size (d_1), the ratio of the coarse pore size to the fine pore size (d_2/d_1), and the ratio of the coarse pore volume to the fine pore volume (V_2/V_1). The generation procedure of the new artificial algorithm is shown in **Figure 15**.

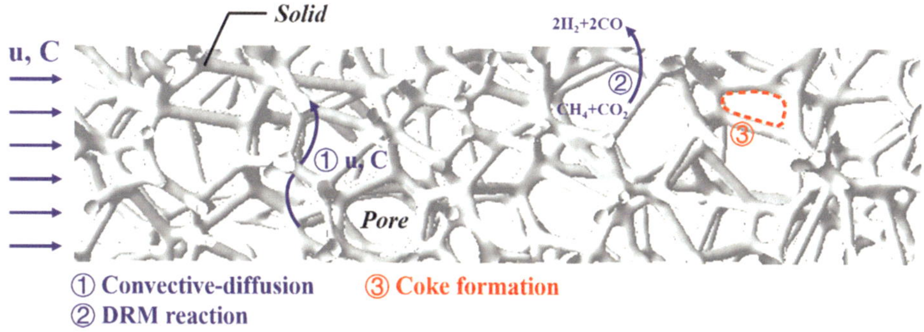

Figure 14.
Schematic diagram of the DRM process in open-cell foam.

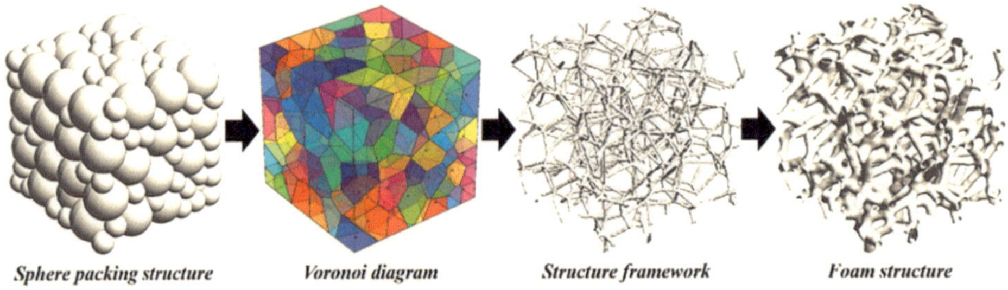

Figure 15.
Schematic diagram of the generation procedure of open-cell foam [39].

3.2 Permeability in open-cell foam with hierarchical pore structure

3.2.1 Effect of hierarchical pore volume ratio

The impact of the hierarchical pore volume ratio (V_2/V_1) on Darcy permeability was studied to elucidate the fluid flow characteristics in open-cell foam with dual-porosity. **Figure 16** compares four hierarchical pore volume ratios (V_2/V_1 = 0, 0.25, 1, and 4) with porosity ranging from 0.76 to 0.95. It is evident that, at the same total porosity, three open-cell foams with hierarchical pore structures (V_2/V_1 = 0.25, 1, 4) exhibit higher permeability than that with uniform pore structure (V_2/V_1 = 0). This result is in line with the experimental findings by Durmus et al. [41], who developed open-cell foam with bimodal pore size. Their results showed that under almost the same porosity, the hierarchical pore structure formed by adding coarse pores to fine pores exhibits lower fluid flow resistance than open-cell foam with only fine pores, which validates the numerical results. Moreover, the increase in V_2/V_1 under the same conditions results in a substantial enhancement of permeability. As shown in **Figure 16**, permeability in the hierarchical pore structure with V_2/V_1 = 4 is more than two times higher than that with V_2/V_1 = 0.25 when porosity is 0.95. This can be explained by the fact that fluid flow resistance in the coarse pore is lower than that in the fine pore, and it decreases further with the increase of V_2/V_1. Notably, permeability increases sharply when $V_2/V_1 > 1$, indicating that fluid flow in the coarse pore dominates the overall permeability.

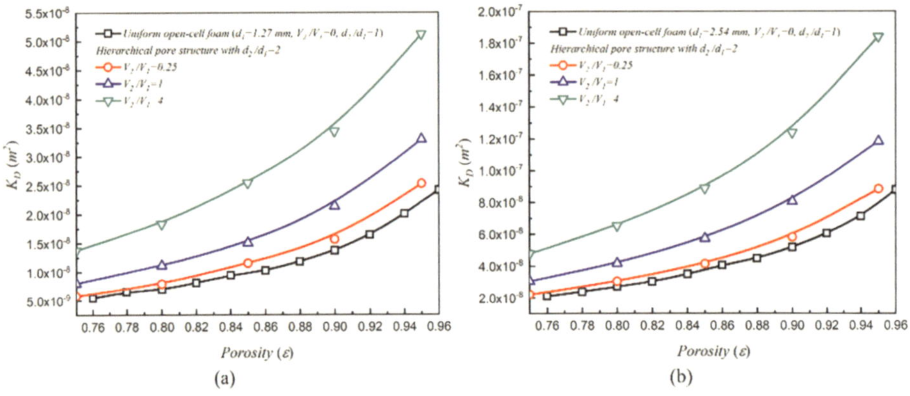

Figure 16.
Effect of V_2/V_1 on permeability [40]: (a) d_1 =1.27 mm, d_2/d_1 =2; (b) d_1 =2.54 mm, d_2/d_1 =2.

3.2.2 Effect of hierarchical pore size ratio

Pore size plays a critical factor in influencing the mass transfer process in porous media. According to the findings emphasized by Ahmad et al. [42], γ-alumina membranes featuring hierarchical pore structures demonstrate reduced transport resistance in comparison to membranes with uniform pore structures. As shown in **Figure 17**, this section examined three hierarchical pore size ratios (d_2/d_1 = 1, 2, 4) with porosity ranging from 0.76 to 0.95 to investigate the effect of the hierarchical pore size ratio (d_2/d_1) on permeability. It has been observed that when the diameter of the fine pores (d_1) remains constant, permeability in open-cell foam increases as the diameter of the coarse pores (d_2) increases. Interestingly, the effect of increasing d_2 on permeability is not significant when V_2/V_1 is not larger than 1. This can be explained by the fact that at this stage, the resistance to fluid flow in V_2 is still comparable to that in V_1. Consequently, increasing the coarse pore diameter does not necessarily lead to a clear increase in permeability. However, when $V_2/V_1 > 1$, permeability rises significantly. This occurs because the coarse pores in open-cell foam can effectively connect with each other, and the contribution of the fine pore to permeability becomes negligible. As discussed in the previous subsection, the fluid flow resistance in V_2 is lower than that in V_1. When $V_1 > V_2$, the impact of the fine pore on permeability diminishes, and the increase in permeability becomes more noticeable due to the dominant role of the interconnected coarse pores.

3.3 DRM reaction in open-cell foam with hierarchical pore structure

To gain insight into the coke resistance during DRM, an additional investigation was conducted. The following parameters are computed to capture the reaction performance in DRM.

$$R_{DRM} = \frac{\int_0^S r_1 dS}{V} \quad (4)$$

$$R_{RWGS} = \frac{\int_0^S r_2 dS}{V} \quad (5)$$

$$R_{Coking} = \frac{\int_0^S r_{Coking} dS}{V} \quad (6)$$

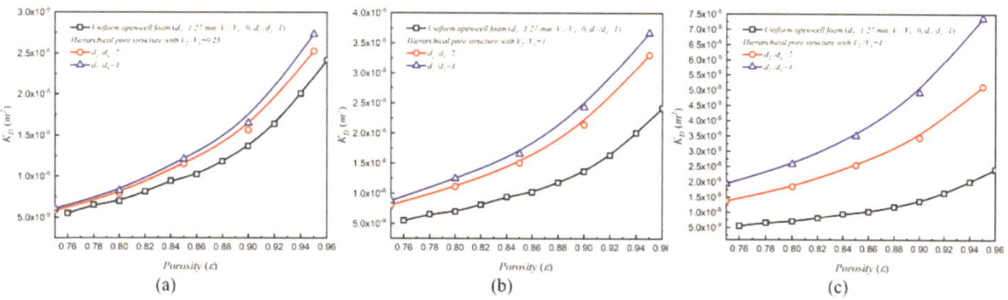

Figure 17.
Effect of d_2/d_1 on permeability [40]: (a) d_1 =1.27 mm, V_2/V_1 =0.25; (b) d_1 =1.27 mm, V_2/V_1 =1; (c) d_1 =1.27 mm, V_2/V_1 =4.

$$S_p = \frac{S}{V} \tag{7}$$

where S_p, R_{DRM}, R_{RWGS}, and R_{Coking} are specific surface area, volume-averaged reaction rates of DRM, reverse water–gas shift, and coke formation, respectively. Within this subsection, fine pore diameter and porosity are defined as 1.27 mm and ε = 0.9, respectively. The impacts of hierarchical pore structures with different V_2/V_1 and d_2/d_1 on initial coke formation are explored.

3.3.1 Effect of hierarchical pore volume ratio

According to Lakiss et al. [14], hierarchical pore structures have the potential to reduce coke formation significantly by providing shorter pathways for mass transport. To gain a more profound understanding of the impact of hierarchical pore structures on coke formation, the distribution of hierarchical pore volume in open-cell foam was taken into account during the investigation. Within this section, the porosity is held constant (ε = 0.9), seven hierarchical pore volume ratios (V_2/V_1) were explored, which are V_2/V_1 = 0, 0.25, 0.5,1, 2, 3, and 4. It is important to note that the hierarchical pore size ratio was kept constant (d_2/d_1 = 2) for this part focusing on hierarchical pore structure ($V_2/V_1 \neq 0$). As depicted in **Figure 18**, the coke formation rates (R_{Coking}) in hierarchical pore structures ($V_2/V_1 \neq 0$) are consistently lower than those in uniform pore structures (V_2/V_1 = 0). Moreover, the coke formation rates (R_{Coking}) exhibit a noticeable downward trend with the increase of V_2/V_1, while the RWGS reaction rate (R_{RWGS}) shows a slight increase, and the DRM reaction rate (R_{DRM}) remains almost stable. This phenomenon can be ascribed to the fact that the resistance to mass transport in V_2 is comparatively smaller than that in V_1. Consequently, increasing V_2/V_1 significantly enhances mass transport performance. From the results shown in **Figure 18**, it becomes evident that the value of R_{Coking} decreases by approximately 33.6% when V_2/V_1 =4, in comparison to the uniform pore structure (V_2/V_1 = 0). **Table 2** reveals that the specific surface area (S_p) follows a decreasing trend with the increase of V_2/V_1. This decrease in active sites accounts for the substantial inhibition of coke formation. However, the progressively diminishing

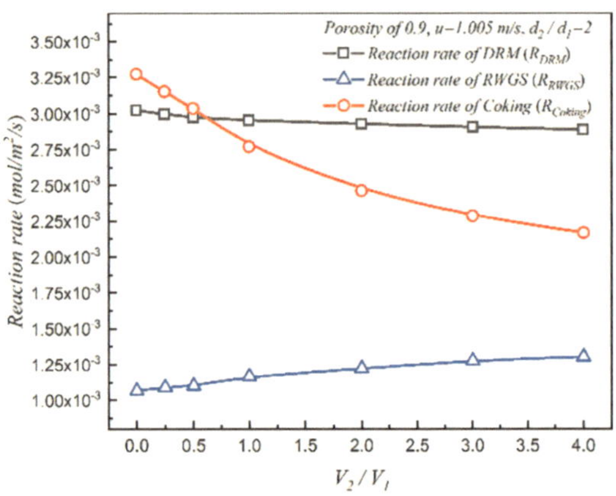

Figure 18.
Effect of V_2/V_1 on reaction rate for DRM [40].

d_2/d_1	1	2	2	2	2	2	2
V_2/V_1	0	0.25	0.5	1	2	3	4
S_p (m^2/m^3)	1984.51	1920.73	1933.73	1671.48	1476.20	1337.77	1300.20

Table 2.
The value of S_p for open-cell foam with different V_2/V_1 [40].

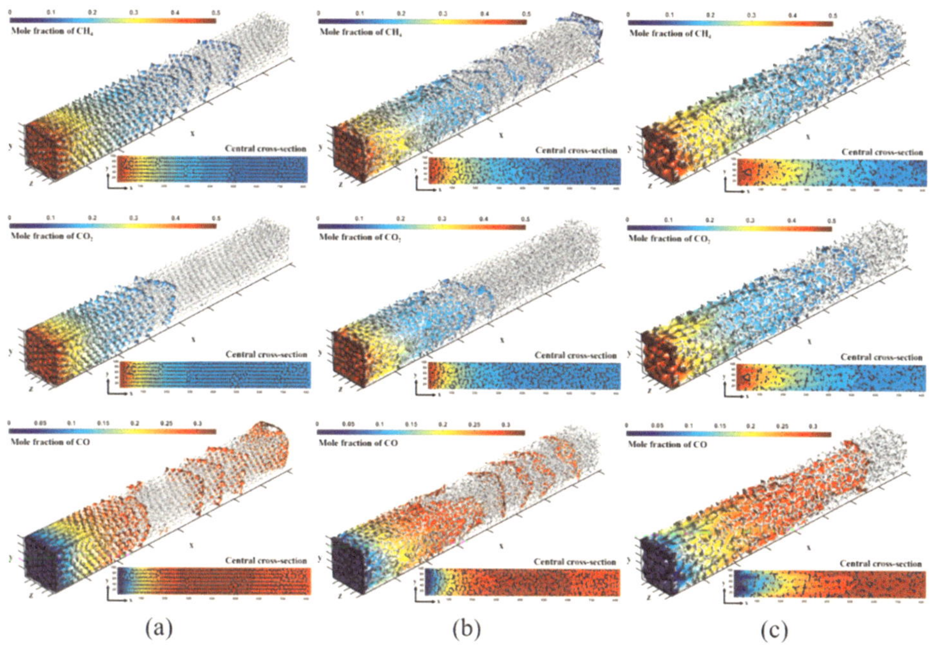

Figure 19.
CH_4, CO_2, *and CO mole fraction distributions with three V_2/V_1 [40]: (a) V_2/V_1 =0, d_2/d_1 =1; (b) V_2/V_1 =0.25, d_2/d_1 =2; (c) V_2/V_1 =4, d_2/d_1 =2.*

quantity of active sites has only a negligible impact on R_{DRM}, indicating that increasing V_2/V_1 not only improves coke resistance but also enhances the catalyst performance.

The characterization of mass transport efficiency across various open-cell foam structures is achieved by visually analyzing the distribution of components along the primary transportation axis. In **Figure 19**, the distributions of mole fractions for CH_4, CO_2, and CO are displayed for three different open-cell foam structures, namely V_2/V_1 = 0, V_2/V_1 = 0.25, and V_2/V_1 = 4. Notably, hierarchical pore structures (V_2/V_1 = 0.25 and V_2/V_1 = 4) exhibit enhanced mass transport efficiency compared to the uniform pore structure (V_2/V_1 = 0). Furthermore, as the hierarchical pore volume ratio (V_2/V_1) increases, the mole fractions of CH_4, CO_2, and CO distribute more uniformly throughout the computational domain, which promotes the coke resistance.

3.3.2 Effect of hierarchical pore size ratio

In this section, the effect of the hierarchical pore size ratio (d_2/d_1) on the coke formation rate in DRM was thoroughly investigated. Based on our prior study, it was found that increasing V_2/V_1 significantly reduces the fluid flow resistance and coke formation. Consequently, the value of V_2/V_1 was set to 4 for the hierarchical pore structure in this analysis. **Figure 20** illustrates the exploration of seven hierarchical pore size ratios, namely d_2/d_1 = 1, 1.5, 2, 2.5, 3, 3.5, and 4. Observing from **Figure 20**, it becomes evident that both R_{DRM} and R_{Coking} gradually decrease as d_2/d_1 increases. However, the decrease rate of R_{Coking} is notably more significant than that of R_{DRM}. Additionally, R_{RWGS} exhibits a slight increase due to the reduction in R_{DRM}, potentially diminishing the selectivity of the DRM reaction. Reviewing the data in **Table 3**, it is observed that the specific surface area (S_p) experiences a sharp decline with the increase of d_2/d_1. Specifically, the specific surface area (S_p) decreases nearly twice from d_2/d_1 = 1 (uniform pore structure) to d_2/d_1 = 4. This substantial reduction in S_p can explain the decreases in R_{DRM} and R_{Coking} to a significant extent. However, it is essential to highlight that R_{Coking} is more sensitive to changes. In **Figure 20**, the transition from d_2/d_1 =1 (uniform pore structure) to d_2/d_1 = 4 results in approximately 13.90% decrease in R_{DRM} but a significant 57.49% decrease in R_{Coking}. This exemplifies that increasing the hierarchical pore size ratio is a potent strategy for promoting coke resistance within in DRM.

In **Figure 21**, the mole fraction distributions for CH_4, CO_2 and CO are presented for three different hierarchical pore size ratios, namely d_2/d_1 = 1, d_2/d_1 = 1.5, and d_2/d_1 = 4. Upon examining **Figure 21**, it becomes evident that, under the constraint of constant V_2/V_1, the component distribution in the open-cell foam structure with d_2/d_1 = 4 is the most uniform throughout the computational domain when compared to the other two structures. This enhanced uniform distribution implies better mass transfer performance, which can effectively promote coke resistance.

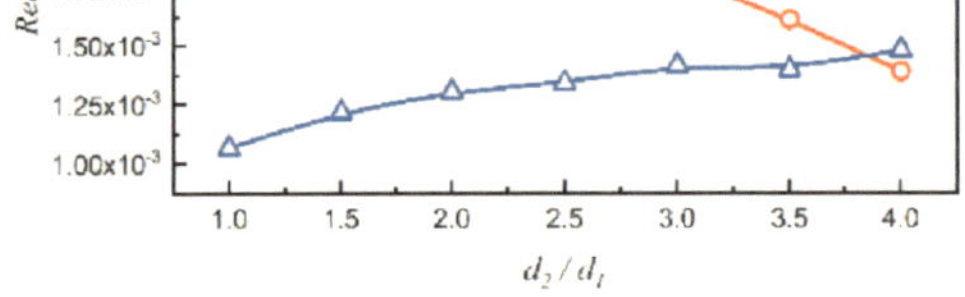

Figure 20.
The effect of d_2/d_1 on reaction rate for DRM [40].

d_2/d_1	1	1.5	2	2.5	3	3.5	4
V_2/V_1	0	4	4	4	4	4	4
S_p (m^2/m^3)	1984.51	1508.59	1300.20	1175.52	1112.11	1001.25	897.44

Table 3.
The value of S_p for open-cell foam with different d_2/d_1 [40].

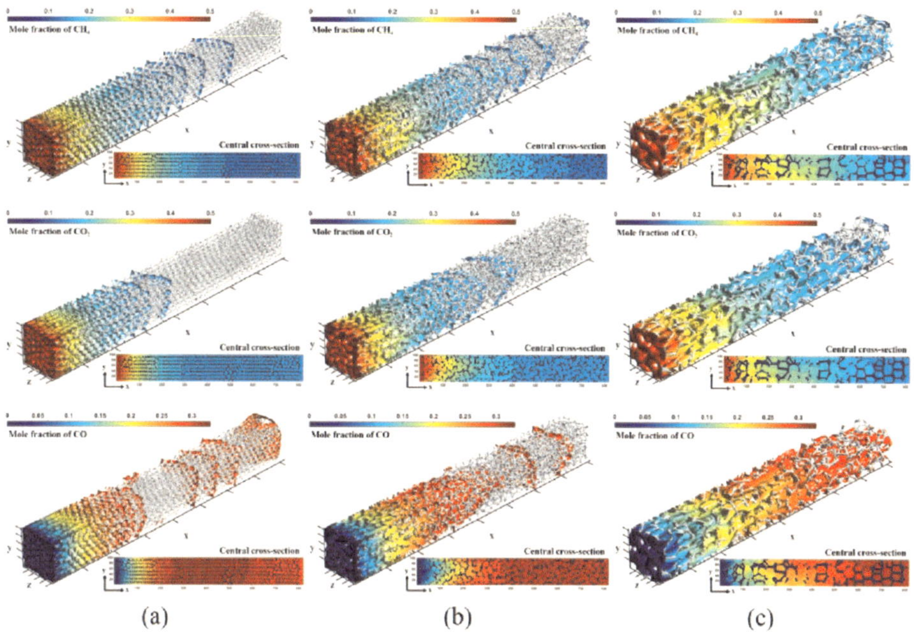

Figure 21.
Mole fraction distributions of CH_4, CO_2, CO *with three* d_2/d_1 *[40]: (a)* $d_2/d_1 = 1$, $V_2/V_1 = 0$; *(b)* $d_2/d_1 = 1.5$, $V_2/V_1 = 4$; *(c)* $d_2/d_1 = 4$, $V_2/V_1 = 4$.

4. Conclusion

To promote the industrial application of DRM, hierarchical pore structure of two porous catalysts, namely porous pellet and open-cell foam, were explored, and two artificial algorithms were proposed to construct hierarchical pore structures for porous pellet and open-cell foam, respectively. The impacts of various hierarchical pore structure parameters on the catalyst performance and coke resistance were investigated.

For porous pellet, the influence of the macro-mesopore structure on intraparticle diffusivity and coke resistance was explored. Under constant reaction conditions (using Ni/Al_2O_3 catalyst, T = 923.15 K, P = 1 bar and F_{CH_4}/F_{CO_2} = 1:1), the catalyst performance is superior when $\varepsilon = 0.7$, $V_{meso}/V_{macro} = 0.5$, and $d_{macro}/d_{meso} = 5$.

For open-cell foam, the impact of the hierarchical pore structure on the fluid flow behavior and coke formation characteristics was examined. The results revealed that increasing the coarse pore volume (V_2) and coarse pore size (d_2) can decrease the fluid flow resistance. Under the restriction of constant hierarchical pore volume ratio ($V_2/V_1 = 4$), increasing d_2/d_1 from 1 (uniform pore structure) to 4 results in a 57.49% decrease in the coke formation rate.

Acknowledgements

The authors would like to express their sincere thanks to the National Natural Science Foundation of China (Nos. 52176062 and 22308058).

Author details

Yixiong Lin, Chen Yang* and Ting Qiu
Fujian Universities Engineering Research Center of Reactive Distillation Technology, College of Chemical Engineering, Fuzhou University, Fuzhou, Fujian, China

*Address all correspondence to: cyang@fzu.edu.cn

References

[1] Wurzel T, Malcus S, Mleczko L. Reaction engineering investigations of CO_2 reforming in a fluidized-bed reactor. Chemical Engineering Science. 2000;**55**:3955-3966

[2] Oyama ST, Hacarlioglu P, Gu Y, Lee D. Dry reforming of methane has no future for hydrogen production: Comparison with steam reforming at high pressure in standard and membrane reactors. International Journal of Hydrogen Energy. 2012;**37**:10444-10450

[3] Guerrero-Caballero J, Kane T, Haidar N, Jalowiecki-Duhamel L, Löfberg A. Ni, Co, Fe supported on ceria and Zr doped ceria as oxygen carriers for chemical looping dry reforming of methane. Catalysis Today. 2019;**333**:251-258

[4] Jang H. A review on dry reforming of methane in aspect of catalytic properties. Catalysis Today. 2019;**324**:15-26

[5] Teh L, Setiabudi H, Timmiati S, Aziz M, Annuar N, Ruslan N. Recent progress in ceria-based catalysts for the dry reforming of methane: A review. Chemical Engineering Science. 2021;**242**:116606

[6] Song Y, Ozdemir E, Ramesh S, Adishev A, Subramanian S, Harale A, et al. Dry reforming of methane by stable Ni-Mo nanocatalysts on single-crystalline MgO. Science. 2020;**367**:777-781

[7] Chen S, Zaffran J, Yang B. Descriptor design in the computational screening of Ni-based catalysts with balanced activity and stability for dry reforming of methane reaction. ACS Catalysis. 2020;**10**:3074-3083

[8] Akri M, Zhao S, Li X, Zang K, Zhang T. Atomically dispersed nickel as coke-resistant active sites for methane dry reforming. Nature Communications. 2019;**10**:5181

[9] Rao S, Coppens M-O. Increasing robustness against deactivation of nanoporous catalysts by introducing an optimized hierarchical pore network-application to hydrodemetalation. Chemical Engineering Science. 2012;**83**:66-76

[10] Baiker A, New M, Richarz W. Determination of intraparticle diffusion coefficients in catalyst pellets-a comparative study of measuring methods. Chemical Engineering Science. 1982;**37**:643-656

[11] Ye G, Wang H, Zhou X, Keil FJ, Coppens MO, Yuan W. Optimizing catalyst pore network structure in the presence of deactivation by coking. AICHE Journal. 2019;**65**:e16687

[12] Arami-Niya A, Rufford TE, Zhu Z. Activated carbon monoliths with hierarchical pore structure from tar pitch and coal powder for the adsorption of CO_2, CH_4 and N_2. Carbon. 2016;**103**:115-124

[13] Yue Y, Liu H, Yuan P, Li T, Yu C, Bi H, et al. From natural aluminosilicate minerals to hierarchical ZSM-5 zeolites: A nanoscale depolymerization–reorganization approach. Journal of Catalysis. 2014;**319**:200-210

[14] Lakiss L, Ngoye F, Canaff C, Laforge S, Pouilloux Y, Qin Z, et al. On the remarkable resistance to coke formation of nanometer-sized and hierarchical MFI zeolites during ethanol to hydrocarbons transformation. Journal of Catalysis. 2015;**328**:165-172

[15] Tao Y, Kanoh H, Kaneko K. ZSM-5 monolith of uniform mesoporous

channels. Journal of the American Chemical Society. 2003;**125**:6044-6045

[16] Scheffler F, Schwieger W, Freude D, Liu H, Heyer W, Janowski F. Transformation of porous glass beads into MFI-type containing beads. Microporous and Mesoporous Materials. 2002;**55**:181-191

[17] Dong A, Wang Y, Tang Y, Zhang Y, Ren N, Gao Z. Mechanically stable zeolite monoliths with three-dimensional ordered macropores by the transformation of mesoporous silica spheres. Advanced Materials. 2002;**14**: 1506-1510

[18] Zheng Y, Wang Q, Yang C, Qiu T. Experimental study on mass transport mechanism in poly (styrene-co-divinylbenzene) microspheres with hierarchical pore structure. Chemical Engineering and Processing-Process Intensification. 2019;**139**:183-192

[19] Twigg M, Richardson J. Theory and applications of ceramic foam catalysts. Chemical Engineering Research and Design. 2002;**80**:183-189

[20] Twigg M, Richardson J. Fundamentals and applications of structured ceramic foam catalysts. Industrial & Engineering Chemistry Research. 2007;**46**:4166-4177

[21] Richardson J, Garrait M, Hung J. Carbon dioxide reforming with Rh and Pt-Re catalysts dispersed on ceramic foam supports. Applied Catalysis A: General. 2003;**255**:69-82

[22] Chen X, Wang F, Han Y, Yu R, Cheng Z. Thermochemical storage analysis of the dry reforming of methane in foam solar reactor. Energy Conversion and Management. 2018;**158**: 489-498

[23] Zhang H, Shuai Y, Lougou BG, Jiang B, Yang D, Pan Q, et al. Effects of foam structure on thermochemical characteristics of porous-filled solar reactor. Energy. 2022;**239**:122219

[24] Zhang Z, Zhao G, Bi G, Guo Y, Xie J. Monolithic SiC-foam supported Ni-La_2O_3 composites for dry reforming of methane with enhanced carbon resistance. Fuel Processing Technology. 2021;**212**:106627

[25] Benguerba Y, Dehimi L, Virginie M, Dumas C, Ernst B. Modelling of methane dry reforming over Ni/Al_2O_3 catalyst in a fixed-bed catalytic reactor. Reaction Kinetics, Mechanisms and Catalysis. 2015;**114**:109-119

[26] Liu X, Wang H, Ye G, Zhou X, Keil F. Enhanced performance of catalyst pellets for methane dry reforming by engineering pore network structure. Chemical Engineering Journal. 2019;**373**: 1389-1396

[27] Carvalho D, Ferreira N, Josué Filho M, Ferreira O, Soares J, Oliveira A. Ni-Fe and Co-Fe binary oxides derived from layered double hydroxides and their catalytic evaluation for hydrogen production. Catalysis Today. 2015;**250**: 155-165

[28] Hussain M, Tian E, Cao T-F, Tao W-Q. Pore-scale modeling of effective diffusion coefficient of building materials. International Journal of Heat and Mass Transfer. 2015;**90**:1266-1274

[29] Chen L, Zhang R, Min T, Kang Q, Tao W. Pore-scale study of effects of macroscopic pores and their distributions on reactive transport in hierarchical porous media. Chemical Engineering Journal. 2018;**349**:428-437

[30] Lin Y, Yang C, Wan Z, Qiu T. Lattice Boltzmann simulation of intraparticle

diffusivity in porous pellets with macro-mesopore structure. International Journal of Heat and Mass Transfer. 2019; **138**:1014-1028

[31] Peng Z, Chen L-H, Sun M-H, Zhao H, Wang Z, Li Y, et al. A hierarchical zeolitic Murray material with a mass transfer advantage promotes catalytic efficiency improvement. Inorganic Chemistry Frontiers. 2018;**5**: 2829-2835

[32] Wang J, Li H, Shuang C, Li A, Wang C, Huang Y. Effect of pore structure on adsorption behavior of ibuprofen by magnetic anion exchange resins. Microporous and Mesoporous Materials. 2015;**210**: 94-100

[33] Lin Y, Yang C, Choi C, Zhang W, Machida H, Norinaga K. Lattice Boltzmann simulation of multicomponent reaction-diffusion and coke formation in a catalyst with hierarchical pore structure for dry reforming of methane. Chemical Engineering Science. 2021;**229**:116105

[34] Lu T, Stone H, Ashby M. Heat transfer in open-cell metal foams. Acta Materialia. 1998;**46**:3619-3635

[35] Lacroix M, Nguyen P, Schweich D, Huu CP, Savin-Poncet S, Edouard D. Pressure drop measurements and modeling on SiC foams. Chemical Engineering Science. 2007;**62**:3259-3267

[36] Krishnan S, Murthy JY, Garimella SV. Direct simulation of transport in open-cell metal foam. Journal of Heat Transfer. 2006;**128**(8): 793-799

[37] Richardson J, Peng Y, Remue D. Properties of ceramic foam catalyst supports: Pressure drop. Applied Catalysis A: General. 2000;**204**:19-32

[38] Boomsma K, Poulikakos D, Ventikos Y. Simulations of flow through open cell metal foams using an idealized periodic cell structure. International Journal of Heat and Fluid Flow. 2003;**24**: 825-834

[39] Lin Y, Yang C, Zhang W, Fukumoto K, Saito Y, Machida H, et al. Estimation of effective thermal conductivity in open-cell foam with hierarchical pore structure using lattice Boltzmann method. Applied Thermal Engineering. 2023;**218**:119314

[40] Lin Y, Yang C, Zhang W, Machida H, Norinaga K. Lattice Boltzmann study on the effect of hierarchical pore structure on fluid flow and coke formation characteristics in open-cell foam for dry reforming of methane. Chemical Engineering Science. 2023;**268**:118380

[41] Durmus F, Maiorano L, Molina J. Open-cell aluminum foams with bimodal pore size distributions for emerging thermal management applications. International Journal of Heat and Mass Transfer. 2022;**191**:122852

[42] Ahmad A, Leo C, Abd. Shukor S. Tailoring of a γ-alumina membrane with a bimodal pore size distribution for improved permeability. Journal of the American Ceramic Society. 2008;**91**: 246-251

Chapter 10

Transport Mechanisms in Membranes Used for Desalination Applications

Hesam Bazargan Harandi and Anahita Asadi

Abstract

Membranes have been widely utilized for different applications in various fields such as desalination and separation processes. The membrane-based desalination technologies, accounting for over 65% of desalination plants, share the common objective of extracting freshwater from various feed sources including seawater, high total dissolved solids brine extracted from deep saline aquifers, brackish water, and industrial effluent. However, the transport mechanisms within the membranes of these aforementioned techniques are significantly different in nature. These differences arise from the distinct specifications of membranes used in each technique, as well as the driving forces applied in their respective processes. Therefore, this chapter focuses on analyzing the transport mechanisms involved in the membranes of three common membrane-based desalination processes: reverse osmosis (RO), membrane distillation (MD), and electrodialysis (ED). In addition to the transport mechanisms occurring within the membranes, this chapter examines the driving forces, the required specifications of the membranes, and the performance parameters and energy consumption associated with the processes mentioned above.

Keywords: transport mechanisms, reverse osmosis (RO), membrane distillation (MD), electrodialysis (ED), diffusion, convection, migration, mass/heat transfer

1. Introduction

Due to the growing demand for freshwater across various sectors such as drinking, agriculture, and industries arising from water scarcity occurring globally even in locations currently considered as water-rich regions, addressing these problems has become urgent, calling for a tremendous amount of research for identifying robust solutions [1–3]. Desalination of various water sources, such as seawater, brackish water, high total dissolved solid (TDS) brine extracted from deep saline aquifers, industrial effluent, etc., has emerged as a viable solution for freshwater production. These desalination technologies have witnessed significant advancements in recent decades. In the past decade, there has been a notable shift in focus toward energy-efficient desalination techniques that offer reduced costs and lower energy consumption, garnering increased interest when compared to conventional desalination technologies [4]. In general, conventional desalination methods can be broadly classified into two categories: membrane-based and thermal-based methods. Among the thermal methods, Multi-Stage Flash (MSF) and Multi-Effect Desalination (MED) have emerged as the most prevalent techniques for freshwater production from saltwater, utilizing the evaporation and condensation process. Thermal-based desalination technologies, being known as energy-intensive methods, are mostly common in Gulf regions [1–3, 5].

Membranes, both naturally and synthetically, act as porous mediums and hold significant importance in various fields, from the operations of living organisms to separation technologies. Naturally, the skin, lungs, and kidneys are acknowledged as crucial membranes within the human body [3]. Membrane-based desalination and separation techniques, both conventional and emerging, have gained widespread utilization globally, constituting a majority share of over 65% in desalination plants [6]. Various membrane-based desalination technologies have emerged, employing diverse driving forces, including pressure-driven, temperature-driven, and electrical-driven mechanisms. Reverse osmosis (RO) stands as the leading first-generation membrane-based desalination technology and is commonly utilized in desalination plants worldwide [7]. As a pressure-driven membrane-based technique, RO functions by applying an operating pressure that surpasses the summation of osmosis pressure and friction loss [8, 9]. In contrast to RO, membrane distillation (MD) has emerged as a prominent second-generation membrane-based desalination technology that utilizes the vapor pressure difference, primarily arising from temperature differences in streams, as the driving force for the desalination process [10]. The operational advantages of MD lie not only in its utilization of low-grade heat, which makes it feasible to integrate with renewable energy and waste heat sources but also in its lower sensitivity when effectively treating feed water with high concentrations [11, 12]. These factors contribute to the increased attractiveness of MD systems when compared to other desalination technologies. Electrodialysis (ED) is a practical method for desalination that utilizes an electric field to separate salt and other ions from water [13]. This electrical-driven technique employs a membrane stack comprising alternating anion/cation-exchange membranes, which selectively allow the passage of ions based on their charge.

While all the previously mentioned membrane-based desalination technologies share the common objective of extracting freshwater from the feed water, the transport scenarios within the membranes of these techniques exhibit significant differences. These differences arise from the distinct specifications of membranes used in each technique, as well as the driving forces applied in their respective processes. To develop the performance of the membrane-based desalination technologies, a multidisciplinary endeavor is imperative, centered on fabricating highly suitable membranes with improved specifications as well as optimizing the operating parameters and module geometries. Such a comprehensive approach is essential for fostering progress and innovation in the field of membrane-based technologies for diverse applications such as desalination.

In this regard and based on the results obtained by our research group in the past decade, this chapter comprehensively elucidates the transport mechanisms involved in the membranes employed in three prevalent and emerging membrane-based desalination processes, i.e., RO, MD, and electrical potential difference (ED). Furthermore, this chapter provides a comprehensive and in-depth overview of the aforementioned membrane-based desalination techniques. It delves into the historical background and development of these techniques, the basic technical concept of processes, the desired properties of the membranes utilized, and the key performance parameters associated with them.

2. Reverse osmosis (RO)

2.1 Introduction and historical background

The first generation of membrane-based separation processes, i.e., reverse osmosis (RO), nanofiltration (NF), ultrafiltration (UF), and microfiltration (MF), mainly relies on pressure-driven mechanisms [8]. MF membranes possess the largest pore size among other types, enabling them to effectively only reject large particles and various microorganisms [3, 14]. UF membranes have smaller pores compared to MF membranes. Consequently, UF membranes not only reject large particles and microorganisms but also demonstrate the capability to exclude bacteria and soluble macromolecules such as proteins [8]. NF membranes, characterized by porous structures, exhibit enhanced separation performance in comparison with UF membranes due to their small pore sizes being on the order of 1 nm. Lastly, RO membranes excel at excluding particles, as well as many low molar mass species such as salt ions and organics. The figure presented below illustrates the spectrum of nominal membrane pore sizes (**Figure 1**) [8].

There are two inherent distinctions between RO and other filtration processes, namely, MF, UF, and NF. First, filtration processes rely on a sieving mechanism for separation, wherein the membrane allows the passage of smaller particles while retaining larger ones [3]. Conversely, RO membranes selectively permit the permeation of the solvent solely, effectively retaining the solute. As the second distinction, MF, UF, and NF processes are employed for the separation of suspended materials, whereas RO is utilized specifically for the separation of dissolved solids. Indeed, RO entails the use of a selectively permeable (semipermeable) membrane with the capability to effectively eliminate all low-molecular-weight organic compounds as well as monovalent and multivalent ions through rejection. While MF, UF, and NF are highly efficient as pretreatment stages, the increasingly stringent standards for purified water necessitate the prominent utilization of RO in the desalination of feed water [15]. RO, which is currently employed in almost 60% of desalination plants worldwide, has a rich historical background that turned it into the most predominant and commercially established technique for freshwater production [3]. The conceptualization of RO originated in the early 1950s at the University of California, Los Angeles (UCLA), and was spearheaded by Professor Samuel Yuster [16]. Professor Yuster's idea involved employing Gibbs' adsorption equation as a framework to explore methods for generating freshwater from brackish and seawater sources. According to Gibb's equation, a distinct layer of relatively pure water should exist at the interface of brine and air, or any other hydrophobic surface, thereby presenting an opportunity to extract freshwater [16]. The initial concept aimed to allow the formation of pure water

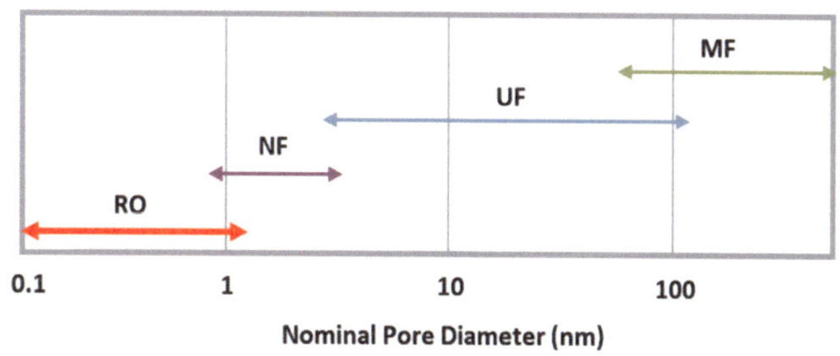

Figure 1.
Range of nominal pore diameter for first-generation membrane-based separation processes.

at the solid-liquid interface and subsequently harvest this water by compelling it to flow through the pores of a solid material under high pressure. In 1958, a groundbreaking milestone was achieved with the successful implementation of a flat plastic film, supported by a porous plate, in an experimental setup at UCLA [16–18]. Coincidentally, during the same period, independent efforts were underway by Breton and Reid at the University of Florida, who achieved the first successful RO tests using cellulose acetate membranes for seawater [19]. The commercial viability of RO took a significant leap forward in 1958 when Loeb and Sourirajan developed the first asymmetric cellulose acetate membrane, thereby revolutionizing the field [20]. Following this achievement, a substantial body of research studies was performed, centering on the development of novel membrane materials and the comprehensive evaluation of their performance. These investigations have significantly contributed to the advancement of RO technology and its application in desalination. In the mid-1970s, the introduction of cellulose triacetate hollow fiber membranes by Dow Chemical Company marked a significant milestone in the field. This achievement resulted in a remarkable enhancement in the produced flux of RO systems. In the past five decades, RO commercially stands as the most favored and widely adopted membrane-based technology for seawater desalination, primarily owing to its notably lower energy consumption rate. Extensive research and studies have consistently demonstrated that RO processes typically necessitate an energy input ranging from 4 to 6 kWh_{elec}/m^3 for the treatment of seawater, while it ranges from 1.5 to 2.5 kWh_{elec}/m^3 for brackish water [12, 21, 22]. These findings highlight the energy efficiency and practicality of RO in addressing the pressing global need for seawater treatment and freshwater production. Although RO technology is known for its lower energy consumption compared to other desalination methods, it is not without its challenges. Treating high TDS feed water using RO can still encounter significant issues, particularly related to scaling and fouling. These concerns pose important considerations in the successful application of RO for such brine treatment processes.

2.2 RO basic concept and process

Figure 2 illustrates a schematic representation of the osmosis and RO phenomena. As illustrated in **Figure 2**, the direction of solvent flow is determined by its chemical

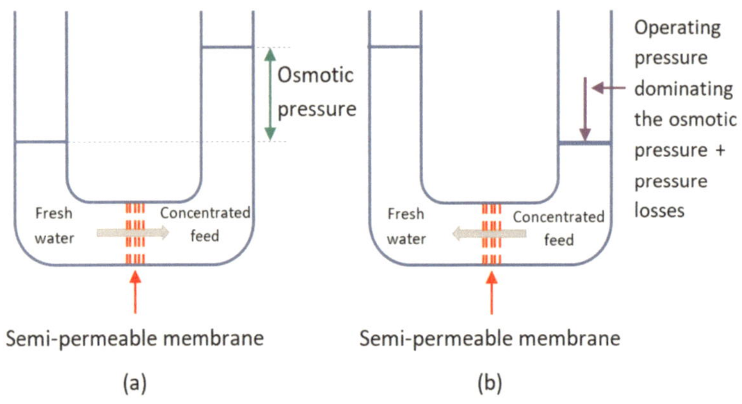

Figure 2.
Schematic representation of: (a) osmosis phenomena and (b) RO process.

potential, which depends on factors such as pressure, temperature, and the concentration of dissolved solids. When pure water is in contact with an ideal semipermeable membrane on both sides, under equal pressure and temperature conditions, there is no net flux of water across the membrane because the chemical potential is balanced on both sides. However, when a soluble salt is introduced on one side, the presence of the salt solution reduces the chemical potential of the water. As a result, osmotic flow occurs from the pure water side to the salt solution side (**Figure 2(a)**), with the aim of restoring the equilibrium of chemical potential [3, 8, 9].

It is worth noting that osmotic pressure is a property of the solution itself and remains independent of the membrane properties. Alternatively, the establishment of equilibrium can also be achieved by applying external pressure, which is equal to the osmotic pressure, to the concentrated feed side. Any additional pressure beyond this point will elevate the chemical potential of the concentrated feed, prompting solvent flow toward the pure water side due to the lower chemical potential it possesses. In fact, RO utilizes the semipermeable membrane to effectively eliminate all low-molecular-weight organic compounds as well as monovalent and multivalent ions. Overall, understanding these principles is crucial for comprehending the intricacies of osmosis and RO processes.

2.3 RO membranes

2.3.1 Membrane materials

The synthetic organic polymers are the main materials used for the fabrication of RO membranes. Generally, the polymers used in RO membranes are either cellulose acetate (CA) or polysulfone coated with aromatic polyamides [8, 9, 14]. Cellulose-based membranes initially held the dominant position in the market for RO membranes. These membranes, which often consist of a rather dense layer of polymer on the surface of an increasingly porous layer, exhibit homogeneity in terms of chemical composition but not in their structural arrangement. CA membranes are crafted from acetylated cellulose, a naturally occurring polymer abundant in plants like cotton. The degree of acetylation signifies the extent to which acetyl groups (CH3COO) replace the hydroxyl (OH) groups dangling from the cellulose chain, in the realm of commercial CA membranes used for RO, a typical degree of acetylation hovers around 2.7. CA membranes have several advantages that distinguish them from other RO membranes currently available. They are characterized by their relatively straightforward manufacturing process and exceptional mechanical properties. Additionally, CA membranes exhibit notable resilience against chlorine-induced deterioration. These membranes can withstand chlorine concentrations of up to 5 ppm, a remarkable threshold that surpasses the tolerance levels of other membrane types, such as those based on aromatic polyamides. Based on aromatic polyamides, thin film composite (TFC) membranes have higher flux and salt rejection in comparison with CA membranes. Most TFC membranes are made with a porous, highly permeable support such as polysulfone, which is coated with a cross-linked aromatic polyamide thin film. The coating provides the salt rejection properties of the membrane. TFC membranes have chemical and structural heterogeneity. In addition to high rejection and high flux, TFC membranes offer other advantages over CA membranes. For one, they can also reject some low-molecular-weight organics. They are also stable over a larger pH range and at higher temperatures than CA.

However, one drawback of TFC membranes is their sensitivity to chlorine. TFC membranes are highly susceptible to attack by chlorine [8].

In addition to organic materials, membranes can also be fabricated using inorganic materials like ceramics or metals. Ceramic membranes, known for their microporous structure, exhibit excellent thermal stability and chemical resistance. However, their widespread adoption has been hindered due to certain drawbacks, including high cost and mechanical fragility, which limit their extensive use [8].

2.3.2 Membrane fabrication methods

In contrast to isotropic membranes, which are typically employed in MF and UF processes and have pores ranging between 50 and 150 nm, RO membranes are commonly anisotropic. Anisotropic membranes are mostly fabricated using phase separation and thin film composite (TPC) methods [8]. CA membranes are generally fabricated by casting a film from a solution of polymer and solvent and immersing the cast film in a nonsolvent for the polymer. TFC membranes are typically composed of a porous substrate that is coated with a thin, impermeable film made from a different polymer. The fabrication of these membranes can be achieved through various techniques, including interfacial polymerization (IP) as the most widely used approach, as well as techniques such as solution coating, plasma polymerization, and surface treatment. In general, the polymerization reaction occurs at the interface of two immiscible liquid phases. To create an extremely thin active layer of polyamide (PA) on top of a supporting membrane, a typical approach involves immersing the substrate into an aqueous solution containing amine monomers (usually ranging from 0.1 to 1% w/v). Subsequently, the substrate is immersed in a second organic solution containing acyl chloride monomers (typically ranging from 0.05 to 0.2% w/v). To enhance the polymerization properties of the PA layer and improve its adhesion to the support membrane surface, the membrane is subjected to heat treatment at temperatures between 70 and 90°C. The IP technique offers significant advantages in independently optimizing the characteristics of the skin layer and the microporous substrate layer.

2.4 Transport mechanisms through RO membranes

To conduct a comprehensive analysis of the transport mechanisms involved in the desalination/separation process, it is crucial to possess a thorough understanding of all potential scenarios of mass transfer that can take place across membranes. It is worth noting that membrane fouling significantly impacts the transport resistances in desalination/separation processes. However, this section investigates the mass transfer occurring through membranes in the absence of fouling, aiming to gain a better understanding of the transport mechanisms involved. There are typically two scenarios that account for transport across the membranes in desalination/separation processes: diffusion through the membrane and pressure-driven convection through pores. These scenarios can be mathematically described by the following equation [8, 9].

$$\vec{J}_i = \rho_i \vec{v} - D_{im} \frac{\vec{\nabla} \rho_i}{l} \tag{1}$$

where J_i [kg/(m^2.s)], ρ_i[kg/m^3], v [m/s], D_{im} [m^2/s], l [m], and $\nabla\rho_i$ [kg/m^3] are the mass flux of component i through the membrane, density of component i, velocity of solution, effective diffusion coefficient of component i in the membrane, characteristic length, and density gradient, respectively. In cases where pore flow plays a substantial role in flux, it is common to employ Darcy's Law as a means to calculate the mass average velocity [8, 23].

$$\vec{v} = -\frac{k}{\mu l}\left(\vec{\nabla} p - \rho \vec{g}\right) \tag{2}$$

here, k [m^2], μ [Pa.s], p [Pa], ρ [kg/m^3], and g [m/s^2] refer to the Darcy permeability constant, solution viscosity, pressure, solution density, and gravity, respectively. By incorporating Eq. (2) into Eq. (1), while considering transport solely in the z-direction (which is typically perpendicular to the membrane surface) and disregarding the effects of gravity, we obtain the following expression:

$$J_{i,z} = \frac{\rho_i k}{\mu}\frac{dp}{dz} - D_{im}\frac{d\rho_i}{dz} \tag{3}$$

In Eq. (3), the first term illustrates the mass flux resulting from pressure-driven convection through pores, while the second term represents the flux arising from diffusion. Due to the nonporous nature of RO membranes, the transport of molecules across the membrane is primarily governed by diffusion. Consequently, the flux across the membrane is predominantly regulated by the second term of Eq. (3). The second term, known as Fick's first law, establishes a relationship between diffusive flux and mass density corresponding to the concentration, assuming a state of equilibrium. According to this law, flux occurs from regions of high concentration to regions of low concentration, and its magnitude is directly proportional to the concentration gradient. In this mechanism, water molecules are absorbed into the upstream face of the membrane, diffuse across the membrane following the chemical potential gradient, and then desorb from the downstream face of the membrane. This mechanism of mass transport across membranes is often known as the "solution-diffusion" model. Moving toward a broader perspective that considers mass transport driven by chemical potential gradients instead of solely concentration gradients, the solution-diffusion transport equation for RO can be mathematically expressed as follows [23, 24]:

$$J_w = c_m(\Delta p - \Delta \pi) \tag{4}$$

Where J_w [kg/(m^2.s)], c_m [kg/(Pa.m^2.s)], Δp[Pa], and $\Delta \pi$ [Pa] are water flux through the membrane, membrane coefficient, the hydrostatic operating pressure difference between two sides of membrane, and the osmotic pressure difference between two solutions on both sides of membrane, respectively. A comprehensive introduction to osmotic pressure will be provided in the upcoming section. While several empirical models, such as the Baker and Wijmans model, Paul model, and others, exist to calculate the membrane coefficient, it is commonly accepted to employ the following equation as a means to describe transport in nonporous membranes [24].

$$c_m = \frac{D_w S_w V_m}{RTl} \tag{5}$$

in which, D_w [m^2/s], R [J/(mol.K)], S_w [kg/m^3], V_m [m^3/mol], and T [K] refer to the diffusion coefficient of water in the dense membrane, universal gas constant, water solubility in the membrane, molar volume of water, and temperature, respectively. As already mentioned, l [m] also exhibits the membrane characteristic length, which is equal to the membrane thickness.

It is important to highlight that the underlying mechanism in membrane-based filtration processes, such as MF and UF, differs significantly. As previously mentioned, the fundamental mechanism driving filtration processes like MF and UF is sieving, where molecules larger than the size of the membrane's pores are effectively rejected. In these membrane-based filtration methods having porous membranes, the contribution of diffusion through the pores is generally considered insignificant compared to convection. When considering the convective as the dominant term in Eq. (3) for transport across UF and MF membranes, the Darcy permeability depends on various membrane structure factors, such as porosity and tortuosity, often in a complex manner.

2.5 Performance parameters of RO process

The evaluation of fabricated RO membranes and module configurations is conducted through the assessment of their performance and operating parameters. The performance of the RO process is characterized by several key variables, which are defined in this section.

- Osmotic and operating pressure

The operating pressure, along with the osmotic pressure, plays a crucial role in determining the efficiency and effectiveness of the RO process. These pressures directly determine the driving force for water permeation through the membrane and, as a result, influence the overall separation performance. The osmotic pressure of a feed solution can be determined through experimental measurements of the concentration of dissolved salts in the solution, while keeping the temperature and pressure constant. The following equation is commonly used to quantify the osmotic pressure of a solution [3, 8]:

$$\pi = RT\sum c_i \tag{6}$$

where π [Pa], R [J/(mol.K)], T [K], and c_i [mol/m^3] are osmotic pressure, universal gas constant, temperature, and concentration of each composition in feed water, respectively. To overcome their adverse impacts, the operating pressure in RO is adjusted to be higher than the sum of osmotic pressure, friction losses, membrane resistance, and permeate pressure.

- Salt rejection

The ability of the RO membrane to reject dissolved solutes is a critical parameter for assessing its effectiveness in desalination applications. This parameter is calculated as follows [3]:

$$SR = \left(1 - \frac{C_p}{C_f}\right) \times 100 \tag{7}$$

where SR [%], C_p [mol/m^3], and C_f [mol/m^3] refer to salt rejection, permeate concentration, and concentration of feed solution, respectively. Higher salt rejection rates indicate a membrane's capability to effectively remove more dissolved solutes from the feed solution, resulting in the production of higher-quality purified water.

- Permeate recovery

Permeate recovery refers to the ratio of produced freshwater to the feed solution. It is an important parameter to determine the overall efficiency and sustainability of the RO process. This parameter is defined as follows [3, 8, 15]:

$$PR = \left(\frac{m_p}{m_f}\right) \times 100 \tag{8}$$

in which PR [%], m_p [kg/(m^2.s)], and m_f [kg/(m^2.s)] are permeate recovery, produced permeate flow rate, and feed flow rate, respectively. Higher permeate recovery rates indicate more efficient feed utilization to produce more freshwater.

3. Membrane distillation (MD)

3.1 Introduction and historical background

In response to growing global concerns surrounding water and energy scarcity, there has been a significant increase in the focus on low-energy desalination technologies [25, 26]. One such technology that has garnered considerable attention is membrane distillation (MD), which offers notable advantages not only in desalination and wastewater treatment but also in diverse applications such as industrial recycling, as well as separation processes in the pharmaceutical, medical, and food industries [27, 28]. Unlike most membrane-based separation processes, which are typically isothermal and rely on transmembrane chemical potential differences (such as hydrostatic pressure, concentration, and electrical potential), MD, as a pioneer of the second-generation of membrane-based desalination technologies, operates as a nonisothermal process [11, 25]. The MD process possesses distinct advantages in comparison with alternative processes. MD is driven by the partial pressure difference which can be even supplied by a low-temperature difference. This lower operating temperature can be provided by eco-friendly sources like renewable energy and the harvesting of waste heat [11]. In addition to the advantages of reduced operating temperature, which translates to lower energy costs, and the simplicity of equipment, the process safety of the MD technique is higher than other membrane-based desalination technologies, owing to its lower operating pressure [29, 30]. It is worth noting that the MD has a long-standing history with a significant chronological gap. The first patent for MD systems was introduced by Bodell in 1963 [31]. Subsequently, Weyl filed the second MD patent in the United States in 1967 [32]. His patent claimed an enhanced method and apparatus for extracting freshwater from saline water, utilizing direct contact membrane distillation (DCMD). In this improved approach, two types of water, whether stationary or in motion relative to the membrane, were employed, and the process was conducted in a single stage or through multiple stages. Weyl employed a polytetrafluoroethylene (PTFE) membrane, characterized by a porosity of 42%, a thickness of 3175 μm, and an average pore size of 9 μm. In 1967, Findley conducted experiments using the DCMD configuration and investigated the performance of different membrane materials such as paper hot cup, gumwood, aluminum foil, cellophane, glass fibers, paper plate, diatomaceous earth mat, and nylon [33]. To achieve the necessary hydrophobicity of the membranes, coatings made of silicone and Teflon were employed. Findley analyzed the MD experimental results and identified the membrane characteristics that are most suitable for successful MD operation. Building upon this work, Bodel, in 1968, introduced a novel system and method for converting nonpotable aqueous solutions into potable water [34]. His second MD patent involved the utilization of a parallel array of tubular silicone membranes. Without a break, researchers worldwide introduced and modeled various configurations of MD. Despite the initial rapid progress in MD, the interest in this process waned, gradually losing its allure. This decline can be attributed, in part, to the observed lower produced flux of MD when compared to that of the RO process. In the 1980s, the attention of academic communities was paid back again to the MD systems when novel membranes with better characteristics became available [35, 36]. These membranes, typically composed of materials, such as PTFE, PP, and PVDF, possessed remarkable attributes, including reduced thickness and porosity levels of up to 90%. Given that ongoing efforts are underway to further enhance membrane performance through modifications, MD has emerged as a compelling and energy-efficient desalination technology based on membrane principles.

Based on the available literature [11, 12, 21], MD has been reported to require an average energy input of 1.25 $kWh_{elec,equi}/m^3$, encompassing both thermal and electrical energy, for a large-scale desalination plant. In contrast to the RO process, which relies solely on electrical energy, MD requires the additional input of thermal energy. Remarkably, approximately 90% of the total energy consumption in MD can be attributed to this heat energy demand. To mitigate the expenses associated with MD, consideration can be given to alternative and cost-effective energy sources. For instance, utilizing solar energy, which is readily available and relatively inexpensive, could prove economically advantageous. By harnessing solar energy as the primary energy source and implementing a continuous energy recovery system, the MD process emerges as an economically viable technology. Moreover, the utilization of waste heat as an energy source presents another promising avenue for reducing MD costs in desalination operations. In such cases, the MD process becomes a favorable alternative when compared to the traditional RO process from an economic standpoint.

Moreover, previous research has identified the potential for integrating MD technology into large-scale seawater RO desalination plants, with the MD process effectively operating within the brine stream. This is due to MD low sensitivity to the feed concentration. This integration offers a promising approach to enhance overall desalination efficiency and cost-effectiveness in the field of water treatment and supply.

3.2 MD basic concept and process

As mentioned in the previous section, MD represents a thermally driven separation process employing membranes well-suited for treating aqueous feed solutions. In this method, a highly porous and hydrophobic membrane serves as the barrier between the feed solution and permeate [10, 27]. In all MD configurations, the feed solution is in direct contact with the membrane surface, whereas the positioning of permeate varies across different configurations. The transportation of volatile compounds from the feed channel to the permeate channel in various MD configurations, such as

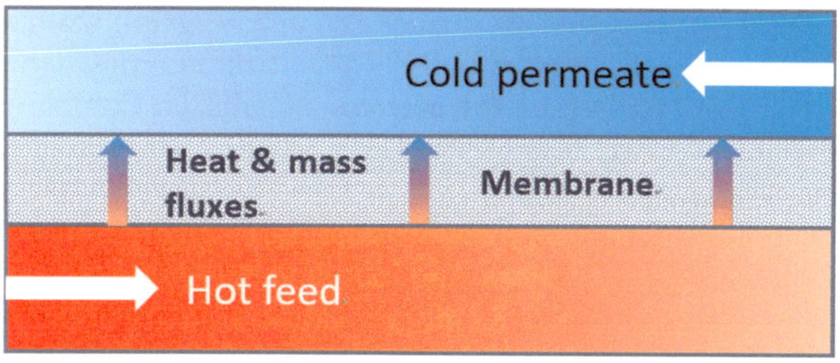

Figure 3.
Principle of heat and mass transfer in MD continuous process.

DCMD, air gap membrane distillation (AGMD), vacuum membrane distillation (VMD), and osmotic membrane distillation (OMD), is primarily driven by the vapor pressure gradient [25, 26]. To establish the required vapor pressure gradient across the membrane, different techniques, like temperature difference as the most common method, are employed in different MD setups. In continuous operations, the feed and permeate aqueous streams, maintained at varying temperature conditions, are tangentially circulated around the MD membrane. Under such temperature conditions, water molecules evaporate at the feed-membrane interface. Subsequently, the vapor diffuses through the membrane pores and condenses at the permeate-membrane interface. **Figure 3** depicts the schematic representation of the MD process for DCMD configuration, consisting of a feed channel, membrane, and permeate channel [25].

The hydrophobic membrane utilized in MD systems serves as a physical barrier, effectively sustaining the formation of liquid-vapor interfaces at the entrances of the membrane pores. This membrane's hydrophobic properties effectively prevent the penetration of liquid water (brine) into the membrane pores from the feed channel. Under the influence of surface tension, the liquid-vapor interface can be maintained up to a certain pressure threshold. Beyond this threshold, liquid water permeates the membrane pores. Termed as the liquid entry pressure (LEP), this pressure can be determined through various experimental tests or theoretically calculated using the Young-Laplace equation [10, 26]:

$$LEP = \frac{-4\sigma \cos\theta}{d_p} \tag{9}$$

In the context of this equation, σ [N/m] denotes the surface tension of the liquid phase, θ[°] represents the contact angle formed between the fluid and the membrane, and d_p [m] signifies the average diameter of the membrane pores.

3.3 MD membranes

3.3.1 Membrane materials

Selecting an optimal membrane for MD applications involves considering factors like low thermal conductivity, high porosity, and minimal tortuosity. Additionally, desirable characteristics include a narrow pore size distribution, superhydrophobic properties with high LEP, longevity, thermal and chemical stability, and an optimized membrane thickness for balancing mass and heat transfer. These criteria collectively contribute to achieving efficient and effective MD performance [11, 25].

Figures 4(a) and **(b)** displays the top surface and cross-section scanning electron microscope (SEM) images of an industrially available membrane, denoted as MS-3010/0.45 μm, manufactured by the Membrane Solution Company [37, 38]. Also, **Figure 4(c)** shows the 3D structure of MS-3010/0.45 μm, generated by the virtual reconstruction approach.

In the fabrication of MD membranes, the materials employed can be broadly categorized into two groups: polymeric and inorganic materials. The frequently utilized polymeric materials for MD membrane fabrication are polytetrafluoroethylene (PTFE), polypropylene (PP), and polyvinylidene fluoride (PVDF), respectively [39–43]. Among the materials introduced above, PVDF has garnered the most attention in research studies due to its exceptional mechanical strength, impressive chemical resistance, favorable thermal stability, and excellent aging resistance. The thermal conductivity of PVDF is measured at 0.17–0.21 W/(m.K). The prevailing method for fabricating PVDF membranes is through phase separation [39, 41, 44]. Alternatively, PTFE, known for its high hydrophobicity, has demonstrated remarkable efficiency as an MD membrane material. It boasts excellent chemical resistance and thermal stability, and its thermal conductivity ranges from approximately 0.25 to 0.29 W/(m.K). However, the fabrication techniques for PTFE membranes are somewhat limited, involving stretching of thin films and sintering of fine PTFE powders [42, 45]. PP, a crystalline polymer, is also characterized by its high hydrophobicity and thermal conductivity of 0.1–0.22 W/(m.K). Nonetheless, its solubility in most solvents at ambient temperatures is minimal. Unfortunately, PP membranes do not possess the same level of hydrophobicity as PTFE, and their fabrication is comparatively more challenging than that of PVDF membranes. As a result, the utilization of PP as a material for MD membranes has been less prevalent compared to PVDF and PTFE [39, 42, 44].

In addition to polymeric materials, mineral substances find utility in the fabrication of MD membranes. Notably, minerals exhibit superior chemical and thermal stability compared to their polymeric counterparts. Indeed, four types of minerals commonly employed for membrane production are metals, zeolites, carbons, and ceramics [46–48]. The mineral membranes are commonly fabricated using the sol-gel process. However, it is essential to acknowledge that ceramic and other inorganic membranes might not be cost-effective under typical thermal and chemical conditions. This aspect warrants further consideration when evaluating their feasibility for specific applications [46–50].

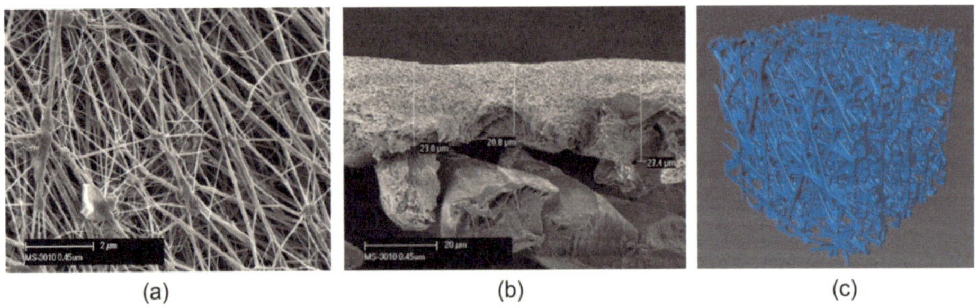

Figure 4.
(a): Top section SEM image; (b): Cross-section SEM image; (c): 3D structure, generated by the virtual reconstruction approach [27], of MS-3010/0.45 μm membrane.

3.3.2 Membrane fabrication methods

The most common techniques employed for the fabrication of MD membranes are sintering, stretching, phase inversion, sol-gel, and electrospinning. Notably, PVDF, PP, and PTFE membranes are commonly manufactured using the phase inversion method, stretching, and either sintering or stretching processes, respectively. Also, sol-gel is the common technique used in the fabrication of mineral membranes, like ceramic membranes. Here, we present concise explanations of these methods widely utilized in membrane production.

- Sintering technique: Powder of polymeric particles is pressed into a film or plate and then sintered just below its melting point [42, 51].

- Stretching technique: In the realm of film production, a significant technique involves the extrusion of a polymer at temperatures near its melting point, alongside a rapid drawdown process. This particular method leads to the alignment of crystallites within the polymer in the direction of the drawing. Subsequently, through annealing and cooling, mechanical stress is applied perpendicular to the aforementioned drawing direction. The outcome of this manufacturing process results in the creation of a remarkably uniform porous structure, characterized by a porosity of approximately 90% [52–54].

- Phase inversion technique: Phase inversion technique involves the dissolution of the polymer in a suitable solvent, followed by casting the resulting solution to a thickness of 20-200 mm onto a suitable support material. The homogeneous solution is then subjected to a coagulant bath, which induces a solvent-nonsolvent exchange, leading to the separation of the solution into two distinct phases: one rich in polymer content and the other rich in liquid content. This process results in the formation of a cast film with desirable properties for various applications [55–57].

- Electrospinning: In this particular technique, an elevated voltage is administered to a polymer solution, prompting the generation of a fine jet that subsequently solidifies into nanofibers [58, 59].

- Sol-gel: These techniques involve the conversion of inorganic precursors into sol (a colloidal suspension) followed by gelation, resulting in the formation of a ceramic matrix. The process allows for precise control over the membrane's composition and microstructure, leading to membranes with enhanced properties for various applications, such as filtration and separation processes [46–50].

3.4 Transport mechanisms through MD membranes

3.4.1 Transmembrane mass transfer theory

In the porous medium of MD, a nonisothermal membrane-based desalination method, the mass transfer flux through the membrane pores is determined by the following equation [4, 12, 25].

$$J_w = c_m\left(P_{mf} - P_{mp}\right) \tag{10}$$

in which J_w [kg/(m^2.s)], c_m [kg/(Pa.m^2.s)], P_{mf} [Pa], and P_{mp} [Pa] refer to the mass transfer flux, mass transfer coefficient in the membrane, water vapor partial pressure at the feed-membrane interface, and water vapor partial pressure at the permeate-membrane interface, respectively. The Antoine equation can be used to calculate the water vapor partial pressures at the local temperatures of the membrane interfaces. Incorporating various transport mechanisms can describe the transportation of volatile compounds through membrane pores in MD. The first effective mechanism is the Maxwell–Stefan mechanism, which corresponds to ordinary molecular diffusion, indicating the transport effect of molecule-molecule collisions. Through this mechanism, water vapor molecules diffuse through the air trapped inside the larger pores of the membrane. The driving force behind this diffusion is the concentration difference between the two interfaces of the membrane. The following is used to calculate the ordinary molecular mass transfer coefficient [4, 25].

$$C_{om} = \frac{1.895e^{-5} \times (T)^{2.072} \varepsilon_m M_w}{RT\tau_m \delta_m \left(\frac{P_{air,mf} - P_{air,mp}}{\ln\left(\frac{P_{air,mf}}{P_{air,mp}}\right)}\right)} \tag{11}$$

where T [K], ε_m, M_w [kg/mol], R [J/(mol.K)], τ_m, and δ_m [m] refer to the temperature, membrane porosity, molecular weight of water, universal gas constant, membrane tortuosity, and membrane thickness, respectively. In addition, $P_{air,mf}$ [Pa] and $P_{air,mp}$ [Pa] are the partial pressures of air at both interfaces of the membrane.

The second mechanism is known as Hagen–Poiseuille, which is characterized by viscous flow. This mode occurs when a total pressure difference exists across the larger pores of the membrane. Through this mechanism, water vapor is transported through these pores *via* viscous flow. The following equation is defined to determine the Hagen–Poiseuille mass transfer coefficient [11, 25].

$$C_{H-P} = 0.03125 \frac{\varepsilon_m d_p^2 P_m M_w}{\tau_m \delta_m RT \mu_v} \tag{12}$$

here, $C_{H\text{-}P}$ [kg/(Pa.m^2.s)], d_p [m], and μ_v [Pa.s] are the Hagen–Poiseuille mass transfer coefficient, mean pore diameter of membrane, and water vapor dynamic viscosity, respectively.

Lastly, the third mechanism is Knudsen diffusion, which occurs within the nanopores of the membrane. This mode comes into play when the characteristic length of the nanopores is comparable to or smaller than the mean free path of water molecules at the given temperature and pressure, indicating the transport effect of molecular-wall collisions. Knudsen diffusion is driven by the partial pressure difference between the two interfaces of the membrane. The Knudsen mass transfer coefficient is mathematically formulated in the following equation [4, 12].

$$C_{kn} = \frac{8}{3} \frac{\varepsilon_m d_m}{2\tau_m \delta_m} \sqrt{\frac{1}{2\pi M_w RT}} \tag{13}$$

Although there are various empirical equations, i.e., the Bruggeman equation, the Millington equation, and Macki–Meares equation, to estimate the pore tortuosity, these correlations cannot be generalized for all kinds of porous medium with various structures. For this reason, according to the membrane structure, it is proposed to employ the following two equations for two types of domains [11]:

a. In loose-packed domains:

$$\tau_m = \frac{1}{\varepsilon_m} \tag{14}$$

b. In distances between closed domains:

$$\tau_m = \frac{(2 - \varepsilon_m)^2}{\varepsilon_m} \tag{15}$$

To assess the prevalence of the aforementioned diffusion mechanisms, ordinary molecular and Knudsen diffusions, the Knudsen number serves as a pertinent indicator for each distinct membrane, considering various specifications like the mean pore diameter. By employing this index, we can effectively compare the dominance of these diffusion mechanisms for a membrane. Knudsen number is defined as follows [10, 25]:

$$Kn = \frac{\lambda}{d_p} \tag{16}$$

in which λ [m] refers to the mean free path of water molecules. At atmospheric pressure, when dealing with the common pore diameters of commercial MD membranes, which typically range between 10 nm and 1 μm, the Knudsen number can be estimated to fall within the range of $0.1 < Kn < 10$. Consequently, one can expect a superposition of both ordinary molecular diffusion and Knudsen diffusion phenomena to occur simultaneously.

Among the models frequently employed in the literature, the dusty gas (DGM), Schofield, and KM-P models stand out as the most widely utilized ones [4, 25, 60]. In general, both models involve three mass transfer mechanisms explained above. As depicted in **Figure 5**, the resistance of ordinary molecular diffusion is in series with the parallel combination of the resistances of Knudsen diffusion and Hagen–Poiseuille (viscous flow) in the Schofield model. In contrast, the series combination of the resistances of Knudsen and ordinary molecular diffusions is in parallel with the resistance of Hagen–Poiseuille (viscous flow) in DGM. As for the KM-P model, the resistance of Hagen–Poiseuille (viscous flow) is placed in series with the parallel combination of the resistances from ordinary molecular and Knudsen diffusions.

3.4.2 Transmembrane heat transfer theory

There are several various heat transfer regions and mechanisms present in the whole module of MD, depending on their particular configuration. However, the overall heat transfer across the membrane occurs through only two mechanisms:

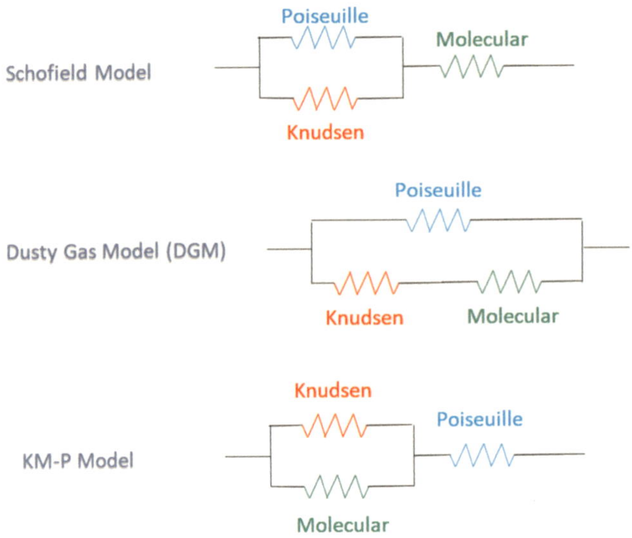

Figure 5.
Commonly adopted models for the prediction of the mass transfer flux across the hydrophobic membrane in MD.

(1) conduction within the solid structure of the membrane and its gas-filled pores and (2) the latent heat carried by water molecules generated during evaporation. The following equation expressed these mechanisms in mathematical representation [4, 12, 25].

$$Q_m = Q_c + Q_v = -\frac{k_m}{\delta_m}\left(T_{mf} - T_{mp}\right) + J_w h_{fg} \tag{17}$$

here, Q_c [W/m^2] refers to the heat loss by thermal conduction, Q_v [W/m^2] is the heat transferred by water vapor flux, k_m [W/(m.K)] stands for the thermal conductivity of the membrane, T_{mf} [K] denotes the temperature at the feed-membrane interface, T_{mp} [K] represents the temperature at the permeate-membrane interface, and h_{fg} [J/kg] corresponds to the enthalpy of water vaporization. The mass transfer flux through the membrane, J_w, is found using transmembrane mass transfer theory, discussed in the previous section. Indeed, the thermal conductivity of a membrane is influenced by both the solid material's thermal conductivity and the gas species' thermal conductivity within its pores. Researchers have proposed several empirical models to calculate the membrane's thermal conductivity. Notably, among these models, the Isostrain, Isostress, and flux low models, as expressed in Eqs. (18)–(20), respectively, have been widely employed in the literature [27].

$$k_m = \varepsilon_m k_w + (1 - \varepsilon_m) k_{Polymer} \tag{18}$$

$$k_m = \left(\frac{\varepsilon_m}{k_w} + \frac{(1 - \varepsilon_m)}{k_{Polymer}}\right)^{-1} \tag{19}$$

$$k_m = k_w \left[\frac{1 + (1 - \varepsilon_m)\beta}{1 - (1 - \varepsilon_m)\beta}\right], \beta = \frac{\frac{k_{Polymer}}{k_w} - 1}{\frac{k_{Polymer}}{k_w} + 2} \tag{20}$$

where k_w [W/(m.K)] and $k_{Polymer}$ [W/(m.K)] are the thermal conductivity of water vapor which pass through the membrane pores and thermal conductivity of polymer, respectively. Based on the findings of combined microstructure and pore-scale modeling for calculating transport properties of MD membranes [27], the Isostrain model is more accurate compared to the Isostress and flux low models for estimating the thermal conductivity of membranes commonly used in MD systems.

3.5 Performance parameters of MD process

This section delves into the performance parameters of the MD system, encompassing both the membrane and its overall efficiency. The thermal efficiency (η), which quantifies the ratio of heat loss to the heat utilized for water vapor production, is expressed by the following equation [4, 12].

$$\eta = \frac{Q_v}{Q_c + Q_v} \tag{21}$$

Q_c [W/m^2] and Q_v [W/m^2] have been already introduced in the previous section. The gain output ratio (GOR) stands as another crucial performance parameter employed to assess and compare different designs of MD modules. Essentially, GOR represents the proportion between the heat generated due to vapor transfer through the membrane and the energy input provided. This metric plays a significant role in evaluating the efficiency and effectiveness of various MD modules [12, 25].

$$GOR = \frac{J_w h_{fg} A_m}{\dot{m}_f C_p \left(T_{fi} - T_{po}\right)} \tag{22}$$

where A_m [m^2], C_p [J/(kg.K)], T_{fi} [K], and T_{po} [K] are the effective membrane area, specific heat capacity at constant pressure, feed inlet temperature, and permeate outlet temperature, respectively.

4. Electrodialysis (ED)

4.1 Introduction and historical background

Electrodialysis (ED) stands as a membrane-based desalination/separation process wherein ions undergo transportation through ion-selective membranes, propelled from one solution to another under the compelling force of an electric field [61]. This intricate transportation phenomenon gives rise to the creation of two distinct solutions: a concentrated one with higher ion content than the original and another comprising nearly pure water. It should be noted that ED only removes ions. Therefore, any bacteria, colloidal material, or silica present in the feed water stream will remain in the product stream. The significance of ED lies notably in its application as an advanced environmental technology, facilitating the development of pristine treatment sequences for water recovery in industrial processes [13, 62].

The timeline of the most important developments for ED is illustrated in **Figure 6**. The concept of ED was first discovered in 1889 by Maigrot and Sabates [63]. They devised an early unit that combined electrolysis and dialysis to eliminate harmful substances that were interfering with sugar manufacturing. In 1890, they successfully demineralized sugar syrup using carbon electrodes, a permanganate paper membrane, and an electric current supplied by a dynamo. The term "electrodialysis" was later

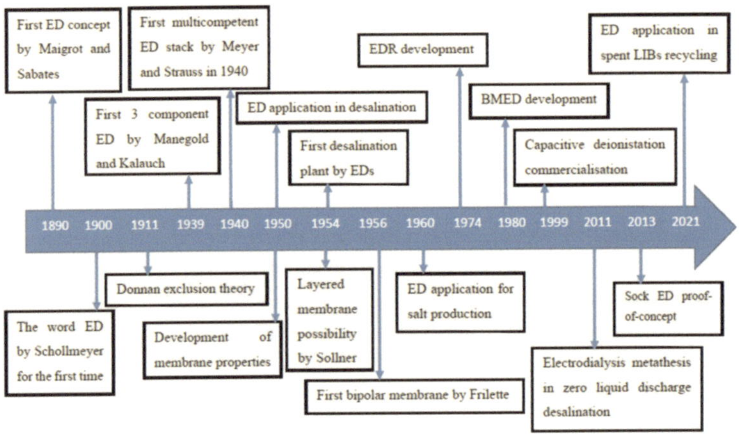

Figure 6.
Timeline of the most important developments for ED.

coined by Schollmeyer around 10 years after these initial experiments [63]. In 1900, Schollmeyer patented the purification of sugar syrup using ED, employing iron as the anode [64, 65]. However, it was not until 1911, when Donnan introduced his exclusion principle, that the theory of ED began to take shape [65]. Building upon the Donnan exclusion principle, the development of polymeric ion-selective membranes started, initially in granule form and gained momentum after 1930 [63, 64]. In 1939, Manegold and Kalauch utilized cation- and anion-selective membranes to construct a 3-compartment ED system [63]. To address energy loss concerns, Meyer and Strauss proposed a significant advancement in 1940—the multicomponent ED stack. This innovative design involved arranging multiple ion-selective membranes in an alternating pattern between the cathode and anode [66]. The significant advancement of ED for desalination and demineralization applications on an industrial scale can be attributed to key developments in membrane technology. In 1950, Juda and McRae [67], followed by Wyllie and Patnode [68], achieved a breakthrough in fabricating low electric resistance, stable, and highly selective membranes. Similarly, Winger et al. [69] at Rohm made crucial contributions in this field in 1953. Prior to 1954, ED had not been employed on an industrial scale. It was not until this year that the inaugural desalination plant, incorporating ED technology, was constructed in Aramco, located in Saudi Arabia [65, 67]. This historical development marked the beginning of ED's utilization in large-scale industrial applications.

Another pivotal moment came with the invention of bipolar membranes, which propelled the progress of ED and paved the way for the emergence of bipolar ED. In 1954, Sollner et al. [70] proposed the possibility of fabricating layered membranes, a concept that sparked new possibilities in the field. Building on this idea, just 2 years later, Frilette [71] successfully developed a bipolar membrane consisting of an anion-exchange membrane (AEM) on one side and a cation-exchange membrane (CEM) on the other side. The concept of electrodialysis reversal (EDR) emerged in 1974 as a promising method for periodically reversing the current, effectively mitigating membrane fouling issues [65, 72]. Subsequently, in 1980, Tomas A. Davis proposed the concept of employing bipolar membranes electrodialysis (BMED) specifically for the production of acids and bases [73]. These breakthroughs in membrane design and technology not only revolutionized ED processes but also opened up new strategies for industrial-scale desalination and demineralization, setting the stage for further advancements in the field, as introduced in **Figure 6**.

The energy demand of ED increases notably with rising feed salinity, making it a high-energy-intensive option compared to other membrane-based desalination methods for concentrated feeds. However, ED proves to be particularly advantageous and favorable when applied to low TDS feed water desalination scenarios. In prior research, it has been established that ED processes typically demand electrical energy ranging from 2.64 to 5.5 kWh_{elec}/m^3 for the treatment of mid-concentrated feed, whereas for low TDS feed, the energy requirement falls within the range of 0.7 to 2.5 kWh_{elec}/m^3. It has been reported that the energy required for desalinating a brine with a TDS concentration of 70,000 ppm was found to be approximately 49.7 kWh_{elec}/m^3, while for a higher TDS brine containing 250,000 ppm, the energy consumption was measured at 175.7 kWh_{elec}/m^3 [21, 61, 62].

4.2 ED basic concept and process

4.2.1 Assembling ED cells and stacks

ED is an electrochemical separation technique that relies on the migration of anions and cations within ion-exchange membranes (IEMs) under an applied electrical field, creating the driving force for ion transport [62]. For this purpose, the AEM and CEM, with positive and negative fixed charges, respectively, are used to selectively allow the penetration of specific ionic species [13]. **Figure 7** depicts an ED cell pair consisting of dilute and concentrate channels, along with AEM, CEM, anode, and cathode. In practical ED units, often referred to as "stacks," several hundred membranes may be

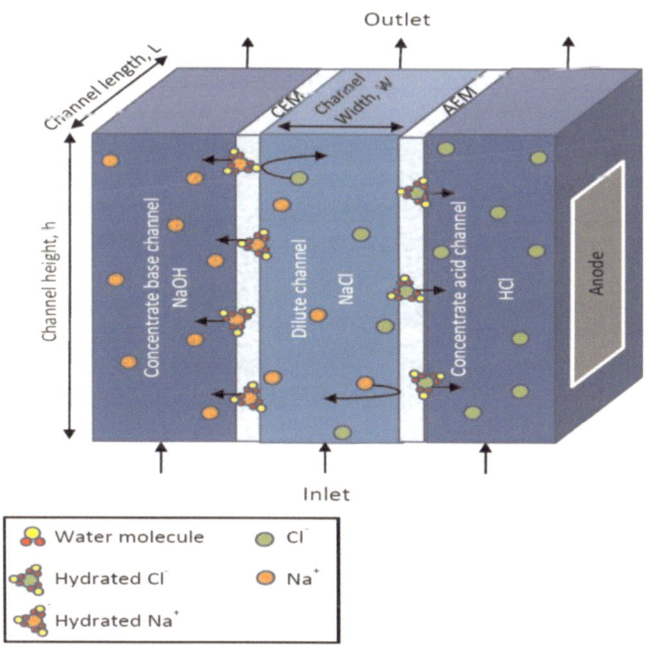

Figure 7.
Schematic representation of an ED process for desalination application.

utilized. The principle behind this stack configuration is that one Faraday passing through a membrane pair can transport one equivalent gram of electrolyte from a diluted compartment to a concentrated one. Hence, incorporating multiple pairs of membranes can enhance the process yield proportionally. When an electrical voltage is applied, cations and anions migrate toward the cathode (negatively charged electrode) and the anode (positively charged electrode), respectively [13]. The cations permeate through the CEM and move into the concentrate base channel, while anions pass through the AEM and enter the concentrate acid channel. In **Figure** 7, the cations and anions of Na^+ and Cl^-, respectively, are employed to depict the ED process.

4.2.2 ED process configurations

The illustration of a single-cell ED system can be found in **Figure** 7 in the previous section. Depending on the various feeding methods employed to supply the inlet streams, single-cell EDs are categorized into two main configurations: single-pass (continuous) and batch setups. Furthermore, the utilization of bipolar membranes enables the integration of multiple EDs into a series arrangement, resulting in what is known as BMED. This innovative approach enhances the capabilities of the ED process.

- Single-pass (continuous) configuration:

In this setup, the outlet solutions from each compartment of the ED unit are collected and stored in separate tanks without being recirculated back into the ED system. Consequently, the concentrations of the inlet solutions remain constant over time and do not undergo any variations during the process. This configuration ensures a continuous flow of solutions through the ED compartments, maintaining a steady state throughout the operation.

- Batch configuration:

Within the batch configuration, a dynamic process occurs, involving the continuous circulation of dilute and concentrate solutions between the ED's channels and their respective tanks. This intricate exchange continues until the desired concentration level is achieved. Throughout this iterative procedure, the inlet solutions' concentrations exhibit temporal variations, adding complexity to the overall system dynamics.

- Bipolar membrane ED (BMED)

In BMED, an innovative approach combines conventional electrodialysis with bipolar membranes to produce acids and bases from corresponding salts, while effectively eliminating unwanted gas production, such as O_2 and H_2. At the heart of BMEDs lies the crucial bipolar membrane, a layered IEM comprising a CEM and an AEM. The success of this technology hinges on several key characteristics, including a high capacity for water dissociation, remarkable stability in highly concentrated acidic and alkaline solutions, superior permselectivity, and minimal electrical resistance.

The water-splitting process within the bipolar membrane, wherein water molecules are divided into proton and hydroxide ions ($2H_2O \rightarrow H_3O^+ + OH^-$), takes place at the interface of the anion and cation permeable layers. This unique arrangement accelerates the generation of hydroxide and proton ions, leading to efficient acid and base production [74]. BMEDs offer distinct advantages over conventional EDs,

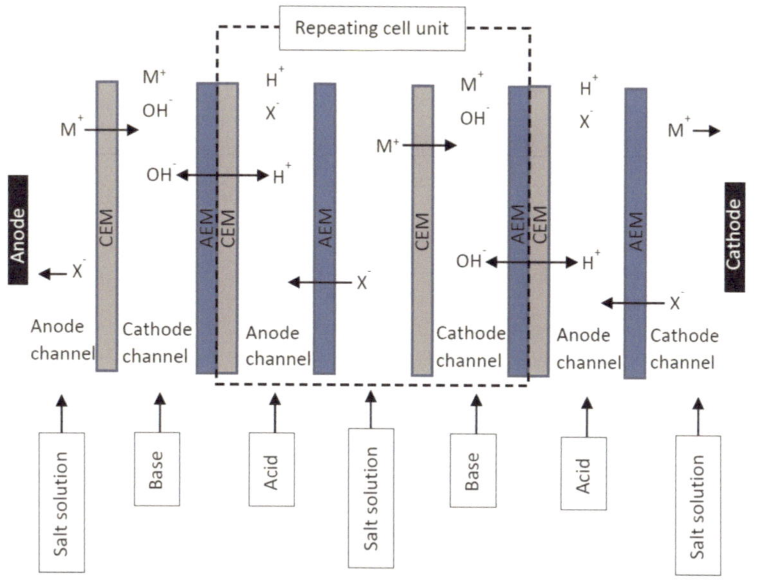

Figure 8.
Schematic representation of BMED stack for acid ad base production.

particularly in terms of lower energy consumption. However, it should be noted that the operational costs of BMEDs are comparatively higher [75, 76]. To provide a visual representation, **Figure 8** depicts the configuration of the BMED stack employed for acids and base production. The stack comprises repeating units, each consisting of a CEM, an AEM, and a bipolar membrane, all arranged between two electrodes. The dilute salt solution enters the central channels of the repeating units, while the production of acidic and alkaline solutions occurs in the anode and cathode channels, respectively.

4.3 ED ion-exchange membranes

4.3.1 Membrane materials

IEMs play a vital role in ED, as sheets of highly swollen ion-exchange resin containing cationic or anionic groups. These membranes are made up of thin polymeric films such as polyethylene, polysulfone, polystyrene, etc., and have charged groups. They exhibit two types of electrical conductivity (charged groups), namely CEMs and AEMs. The AEMs possess a positive charge and are primarily composed of secondary amines ($-NRH_2^+$), tertiary amines ($-NR_2H^+$), quaternary amines ($-NR_3^+$), as well as ammonium groups ($-NH_3^+$, SR_2^+, PR_3^+), all of which are chemically bonded to the polymer membrane's structure. This configuration effectively blocks the passage of positively charged ions while selectively allowing the transit of negatively charged ions. In contrast, the CEMs feature the presence of negatively charged functional groups, including carboxylic acid ($-COO^-$), sulfonic acid ($-SO_3^-$, $C_6H_4O^-$), phosphonic acid ($-PO_3H^-$), and phosphoryl ($-PO_3^{2-}$). These functional groups endow CEMs with the capability to perform functions that are distinctly different from their AEM counterparts [61, 62, 77, 78]. The ions that can freely pass

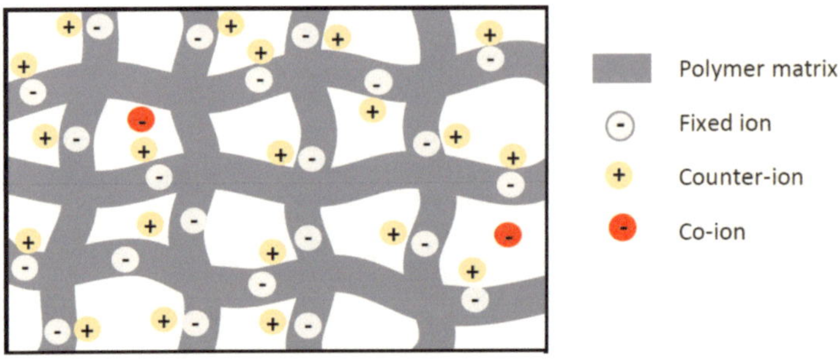

Figure 9.
Schematic representation of a CEM structure.

through CEMs and AEMs, representing positively charged cations and negatively charged anions, are referred to as counter-ions. On the other hand, co-ions, which possess the same charge as the membrane, are impermeable due to the phenomenon of Donnan exclusion [78]. As illustrated in **Figure 9**, the fixed ions remain embedded, while both counter-ions and co-ions are mobile within the polymer matrix. CEMs maintain electrical equilibrium between mobile cations and fixed negative charges, whereas mobile co-ions remain excluded from the polymer matrix. CEMs and AEMs share several key characteristics, including but not limited to low electrical resistance, insolubility in aqueous solutions, semi-rigidity for convenient handling during stack assembly, resistance to osmotic swelling, pH resistance ranging from 1 to 10, long life expectancy, resistance to fouling, and ease of hand-washing.

4.3.2 Membrane fabrication methods

Given the properties of IEMs such as active surface charge distribution, hydrophobicity, resistance to electricity, surface roughness, along with the manufacturing methods of both AEM and CEM, they are classified into homogeneous and heterogeneous IEMs. Homogeneous membranes exhibit a uniform distribution of charge throughout their structure, while heterogeneous membranes contain fixed charges distributed nonuniformly within the polymer matrix. Consequently, homogeneous membranes boast superior electrochemical properties, whereas heterogeneous ones excel in mechanical characteristics. Given the distinct structural and inherent characteristics of these two membrane types, they typically require distinct fabrication methodologies. In the fabrication of homogeneous membranes, common techniques encompass processes such as polymerization, polycondensation of functional monomers, and post-sulfonation of polymers. In addition, heterogeneous membrane preparation entails methods such as dispersing ion-exchange resin within a polymer solution or the fusion and compression of dry ion-exchange resin alongside granulated polymer material.

4.4 Concentration polarization and limiting current density

Electric currents in ED cells are carried by both cations and anions, with similar transfer numbers in aqueous solutions. However, within IEMs, the current transfer is primarily conducted by counter-ions, with a transfer number near unity. The variation in ion transfer numbers between the solution and an IEM results in a decrease in

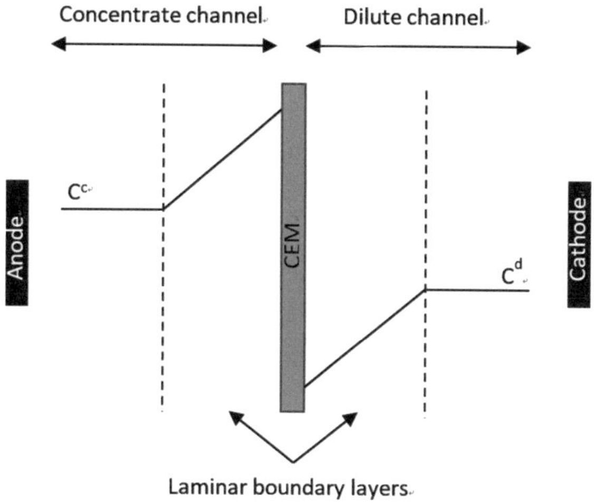

Figure 10.
Concentration polarization at the boundary layer on both sides of a CEM.

electrolyte concentration at the dilute side of the IEM interface, with a corresponding concentration increase on the concentrate side. Consequently, this leads to the establishment of a concentration gradient in the solution between the membrane surface and the bulk, known as concentration polarization. The concentration profile of ions in the dilute and concentration solutions adjacent to the CEM is illustrated in **Figure 10** [79].

According to the theory of concentration polarization, the concentration of ions decreases at the dilute side of the membrane surface. When the electrolyte concentration at the dilute interface approaches zero, there are no more ions available to carry the electric current, resulting in the achievement of the limited current density. This condition corresponds to the increment of voltage drop across the boundary layer, resulting in a higher energy consumption and generation of water dissociation. The consequence of water dissociation is a loss of current utilization and significant pH shifts. This results in a higher pH at the AEM surface and a lower pH at the CEM surface. The pH increase may cause the precipitation of multivalent ions on the membrane, while the pH decrease can harm the membranes themselves. Consequently, in practical ED applications, it is essential to minimize concentration polarization effects and prevent water dissociation [79].

4.5 Transport mechanisms through ED membranes

In electrolyte solutions, the concentrations and velocities of ions vary significantly along the channels. These variations occur due to the continuous flux of ions and water through the IEMs. To comprehend these changes, material balance equations are employed. As the transfer of ions and water molecules transpires from the dilute channel to the concentrate channels, there is a noticeable increase in the concentration and velocity of the solution in the concentrate channels. Conversely, downstream in the outlet of the dilute channel, these quantities experience a decrease. Eq. (23) is used to quantify the transmembrane water flux through the IEMs in all compartments [80].

$$J_w = \frac{\rho_w W(u_{in} - u_{out})}{H} \tag{23}$$

in which, the parameters W [m] and H [m] represent the width and height of the channels. Here, J_w [kg/(m^2.s)] and ρ_w [kg/m^3] denote the transmembrane water flux and density, while u_{in} [m/s] and u_{out} [m/s] refer to the inlet and outlet velocities, respectively. In the context of the dilute channel, J_w represents the collective flow of water molecules passing through both the anion and cation-exchange membranes. The following equation shows the relationship between the transmembrane ion (species) flux through the IEMs and the transmembrane water flux, expressed by Eq. (23) [80].

$$J_i = \frac{W(c_{i,in}u_{in} - c_{i,out}u_{out})}{H} \tag{24}$$

here, J_i [mol/(m^2.s)], $c_{i,in}$ [mol/m^3], and $c_{i,out}$ [mol/m^3] refer to the transmembrane flux of species i, and concentration of species i at the inlet and outlet of channels, respectively.

Various transport mechanisms are responsible for the transportation of ions from solutions through the IEMs. The first effective mechanism is the migration mechanism. Through this mechanism, counter-ions are transferred through the corresponding IEMs, using the electrical field driving force. The second mechanism is diffusion. The driving force behind this diffusion is the ion concentration difference across the IEMs. The convection term constitutes the third transport mechanism involved in the ED process through IEMs. This term arises from two distinct phenomena: osmotic pressure and electro-osmosis. However, it is important to highlight that in ED processes, the influence of osmotic pressure is generally negligible compared to the significant impact of electro-osmotic phenomena. The Nernst-Planck equation is used to explain the roles of the aforementioned transport mechanisms in determining the transmembrane ion flux through IEMs. This equation is mathematically defined as follows [81, 82]:

$$J_i = -z_i M_i^m c_i^m \frac{\Delta\varphi_m}{l_m} - D_i^m \frac{\Delta c_i^m}{l_m} + u_m c_i^m \tag{25}$$

where the first, second, and third terms on the right-hand side are the migration, diffusion, and convection fluxes, respectively. In this equation, z_i, M_i^m [m^2/(V.s)], and c_i^m [mol/m^3] refer to the valence of ions, mobility of species i at membrane medium, and concentration of ion i at IEM interfaces, respectively. Also, $\Delta\varphi_m$ [V], l_m [m], D_i^m [m^2/s], Δc_i^m [mol/m^3], and u_m [m/s], are the voltage difference across IEMs, thickness of the membrane, diffusion coefficient of species i at membrane medium, ion concentration difference across IEMs, and water velocity through IEMs, respectively. The mobility of species i can be estimated using the Nernst-Einstein equation as follows [82]:

$$M_i^m = \frac{FD_i^m}{RT} \tag{26}$$

here, F [96485.33 (A.s)/mol], R [J/(mol.K)], and T [K] are the Faraday's constant, universal gas constant, and temperature, respectively. Then, Faraday's law can be employed to calculate the current density (I) [62].

$$I = F \sum z_i J_i \tag{27}$$

In the literature [83], Eq. (28) is repeatedly suggested to calculate the diffusion coefficients of ions at IEMs by leveraging their diffusivity in the feed solutions in channels.

$$D_i^m = \left(\frac{VF_w}{2 - VF_w} \right)^2 D_i^s \tag{28}$$

in which, VF_w and D_i^s [m^2/s] are volume fraction of pores filled by water in IEMs and ion diffusivity in the feed solutions in channels, respectively. As discussed earlier, the movement of water molecules driven by osmotic pressure across the membrane is found to be ignorable when compared to the dominant electro-osmotic phenomenon. Electro-osmotic water transport occurs due to the transport of hydrated ions under the influence of an electrical potential gradient. Therefore, in the context of determining the transmembrane water velocity mathematically defined in Eq. (29), it is considered that water molecules are primarily drawn into IEMs through the force of electro-osmosis. This force depends on both the applied electrical potential and the friction encountered by ions in the solution [79, 81, 84].

$$u_m = \frac{\varepsilon \zeta}{4 \pi \mu} \frac{\Delta \varphi_m}{l_m} \tag{29}$$

here, ζ [V], ε [F/m], and μ [Pa.s] are the zeta potential, solvent permittivity, and dynamic viscosity, respectively. Given the effective membrane surface area (A_m), the transmembrane water flux, expressed in Eq. (23), is calculated as follows:

$$J_w = A_m u_m \tag{30}$$

The Donnan potential equation describes the potential difference across an interface between an IEM and a solution. This potential difference is influenced by the ion concentrations on both sides of the IEM-solution interface, as well as the pressure difference across the membrane. However, for the specific scenario under consideration, it is assumed that the effect of pressure difference is negligible. Consequently, the simplified form of the Donnan potential equation can be expressed as follows in the context of this study [85–87].

$$\varphi^m - \varphi^s = \frac{RT}{z_i F} \ln \left(\frac{c_i^s}{c_i^m} \right) \tag{31}$$

where superscripts m and s refer to IEM and solution sides of IEM-solution interfaces. Considering the Helmholtz double-layer model [88], the concentration of ion i at IEM interfaces in each channel can be determined as follows:

$$c_i^m = c_i^s \exp \left(\pm \frac{z_i l_i^d}{\varepsilon RT} F^2 (2 r_i) c_i^s \right) \tag{32}$$

in which, positive and negative signs are allocated to AEM and CEM. Also, l_i^d [m] and r_i [m] refer to the thickness of the double layer and radius of counter-ion i moving within the IEM.

4.6 Performance parameters of ED process

The assessment of desalination and treatment efficiency through ED involves the computation of two crucial parameters: percent extraction and current efficiency. These metrics play a pivotal role in evaluating the overall effectiveness of the ED process in purifying and desalinating water.

4.6.1 ED percent extraction

The percentage extraction for each ion in the feed solution for continuous mode is calculated using Eq. (33) [61, 89].

$$PE = \left(\frac{c_{i,out} - c_{i,in}}{c_{i,in}}\right) \times 100 \tag{33}$$

4.6.2 ED current efficiency

In the practical ED processes, it is not possible to fully utilize the current passing through the stack for the purpose of desalinating the feed solution. Besides the electro-osmosis phenomena, there are some other reasons, such as water dissociation, parallel current, inappropriate IEMs, contributing to the current loss. Thus, the total current efficiency denotes the amount of current needed to produce a specific quantity of product water with a designated level of quality. This efficiency is determined by dividing the actual current consumption by the theoretically ideal current required for the process. To express this mathematically, the current efficiency can be defined as follows [61, 62, 90]:

$$CE = \frac{F}{I\Delta t}\left(c_{i,in}u_{in} - c_{i,out}u_{out}\right) \tag{34}$$

where the symbol I represents the electric current density flowing through the system over a specific time interval, denoted as Δt.

5. Conclusion

This chapter offered an in-depth and comprehensive overview of three common membrane-based desalination processes of RO, MD, and ED. As the main purpose of this chapter, a thorough exploration of the transport mechanisms involved in the membranes utilized in the aforementioned membrane-based desalination processes was presented and reviewed. Moreover, this chapter comprehensively presented the historical background and development of these techniques, the basic technical concept of processes, the desired properties of the membranes utilized in each technique, and the key performance parameters associated with them. **Table 1** summarily presents and compares the transport and separation mechanisms occurring within the membranes of each of the aforementioned membrane-based desalination techniques. Furthermore, this table provides a comprehensive analysis of the driving forces, membrane properties, and energy consumption associated with the introduced desalination techniques.

Membrane-based desalination technique	Transport mechanism in membrane	Driving force	Membrane properties	Energy consumption (Electrical equivalent) $kWh_{elec,equi}/m^3$
RO	Diffusion (dominant) + Convection	Total pressure difference (15-80 bar)	Hydrophilic, Dense, High mechanical strength to withstand the higher operational pressure, Resistant to chemical degradation and biological fouling, High fouling resistance, High stability over pH changes, High thermal resistance	1.5-2.5 (BW) 4-6 (SW)
MD	Mass transfer: Ordinary molecular diffusion + Knudsen diffusion + Poiseuille (viscous flow) Heat transfer: Conduction + Latent heat associated with the water vapor molecules	Temperature difference (Vapor pressure difference)	High hydrophobicity, High porosity, High chemical and thermal stability, Low thermal conductivity, Low tortuosity, Low thermal resistance, High scaling resistance	≈1.25
ED	Migration + Diffusion + Convection	Electrical potential difference	Ion-exchange membranes, Low electrical resistance, High fouling resistance, Suitable rigidity, High resistance over pH changes, Long life, High thermal resistance	0.7-2.5 (BW) 2.64-5.5 (MW) ≈49.7 (High TDS)

BW: Brackish water (< 3000 ppm); MW: Moderate water (3000 ppm to 10,000 ppm); SW: Seawater (10,000 ppm to 50,000 ppm); High TDS: Seawater (> 50000 ppm).

Table 1.
Side-by-side assessment of common membrane-based desalination technologies.

RO is a membrane-based separation process driven by the total pressure difference. It is extensively employed for the separation of salts and small organic molecules from liquid streams. The RO membranes typically consist of dense active layers, and the preferred transport mechanisms in these membranes are commonly attributed to two processes: diffusion, which is more dominant and convection. The operating pressure applied in RO, typically ranging from 15 to 80 bar, overcomes the summation of osmosis pressure and friction loss. Despite being a widely adopted and energy-intensive desalination technology, RO does not come without its share of challenges. For instance, when dealing with high TDS feed water, the utilization of RO may still encounter notable issues, primarily concerning scaling and fouling phenomena on the membrane surfaces of RO.

In MD, three mass transport mechanisms of Knudsen diffusion, ordinary molecular diffusion, and Poiseuille (viscous flow) are generally in charge of transporting the water vapor molecules through the membranes. Furthermore, the transfer of heat across the MD membranes transpires through two distinct mechanisms. Firstly, heat is conducted within the solid structure of membrane and its gas-filled pores. Secondly, the latent heat is conveyed through the membrane pores associated with the water vapor molecules. MD membranes should exhibit not only low resistance to mass transfer but also possess high resistance to heat transfer, which aligns with their low thermal conductivity. Moreover, an ideal characteristic of these MD membranes lies in their higher hydrophobicity, which leads to a higher LEP. Additionally, to enhance the effectiveness of MD membranes, it is preferable for them to have lower tortuosity factors and higher porosity. These attributes contribute to a smoother path for mass transfer, thereby enhancing the overall efficiency of the membrane-based desalination process. As shown in **Table 1**, MD stands as a moderately energy-intensive technology for desalination. However, the MD process exhibits a unique advantage as it can effectively utilize low-grade heat as its driving force. This advantage makes MD a promising and competitive technology, especially when renewable energy sources and waste heat are harnessed to provide the necessary thermal energy input.

ED, driven by an electrical potential difference, involves the transportation of ions through ion-selective membranes, which are compelled from one solution to another under the influence of an electric field. In the ED process, both CEM and AEM share common characteristics, such as low electrical resistance, resistance to fouling, solidity to facilitate handling during stack assembly, resistance over pH changes, long life expectancies, suitability for operation in high temperatures, and the convenience of being hand washable. The energy consumption in ED is notably influenced by the concentration of the feed water. This characteristic makes ED a compelling and competitive technology for desalinating brackish water and moderately saline water sources.

Acknowledgements

The authors are grateful for the funding support provided by Key R&D Program of Inner Mongolia Province of China (2021ZD0034) as well as the Science and Technology Plan Project of Sichuan Province (2022YFSY0018). Additionally, Hesam Bazargan Harandi expresses gratitude to Prof. Liwei Zhang for providing the funding for this research.

Nomenclature

J_i	transmembrane flux of species, i mol/(m^2.s) or kg/(m^2.s)
ρ_i	density of species, i kg/m^3
v	velocity of solution, m/s
D_{im}	effective diffusion coefficient of component i in the membrane, m^2/s
$\nabla\rho_i$	density gradient, kg/m^3
k	Darcy permeability constant, m^2
μ	solution viscosity, Pa.s
g	gravitational acceleration, m/s^2
J_w	transmembrane water flux, kg/(m^2.s)
c_m	membrane coefficient, kg/(Pa.m^2.s)
Δp	hydrostatic operating pressure difference between two sides of membrane, Pa
$\Delta\pi$	osmotic pressure difference between two solutions on both sides of membrane, Pa

D_w	diffusion coefficient of water in the dense membrane, m^2/s
R	universal gas constant, J/(mol.K)
S_w	water solubility in the membrane, kg/m^3
V_m	molar volume of water, m^3/mol
T	temperature, K
c_i	concentration of species, i mol/m^3
SR	salt rejection, %
C_p	permeate concentration, mol/m^3
C_f	feed concentration, mol/m^3
PR	permeate recovery, %
m_p	produced permeate flow rate, $kg/(m^2.s)$
m_f	feed flow rate, $kg/(m^2.s)$
σ	surface tension, N/m
θ	contact angle formed between the fluid and the membrane, °
d_p	average diameter of the membrane pores, m
P_{mf}	water vapor partial pressure at the feed-membrane interface, Pa
P_{mp}	water vapor partial pressure at the permeate-membrane interface, Pa
C_{om}	ordinary molecular mass transfer coefficient, $kg/(Pa.m^2.s)$
M_w	molecular weight of water, kg/mol
ε_m	membrane porosity, –
τ_m	membrane tortuosity, –
δ_m	membrane thickness, m
$P_{air,mf}$	partial pressures of air at the feed-membrane interface, Pa
$P_{air,mp}$	partial pressures of air at the permeate-membrane interface, Pa
$C_{H\text{-}P}$	Hagen–Poiseuille mass transfer coefficient, $kg/(Pa.m^2.s)$
μ_v	water vapor dynamic viscosity, Pa.s
C_{Kn}	Knudsen mass transfer coefficient, $kg/(Pa.m^2.s)$
λ	mean free path of water molecules, m
Q_c	heat loss by thermal conduction, W/m^2
Q_v	heat transferred by water vapor flux, W/m^2
k_m	thermal conductivity of the membrane, W/(m.K)
T_{mf}	temperature at the feed-membrane interface, K
T_{mp}	temperature at the permeate-membrane interface, K
k_w	thermal conductivity of water vapor, W/(m.K)
$k_{Polymer}$	thermal conductivity of polymer, W/(m.K)
η	thermal efficiency, –
GOR	gain output ratio, –
A_m	effective membrane area, m^2
C_p	specific heat capacity at constant pressure, J/(kg.K)
T_{fi}	feed inlet temperature, K
T_{po}	permeate outlet temperature, K
W	width of channels, m
H	height of channels, m
u_{in}	inlet velocity, m/s
u_{out}	outlet velocity, m/s
$c_{i,in}$	concentration of species i at the inlet of channels, mol/m^3
$c_{i,out}$	concentration of species i at the outlet of channels, mol/m^3
z_i	valence of ions, –
M_i^m	mobility of species i at membrane medium, $m^2/(V.s)$
c_i^m	concentration of ion i at IEM interfaces, mol/m^3

$\Delta\varphi_m$	voltage difference across IEMs, V
l_m	thickness of the membrane, m
D_i^m	diffusion coefficient of species i at membrane medium, m^2/s
Δc_i^m	ion concentration difference across IEMs, mol/m^3
u_m	average water velocity through IEMs, m/s
F	Faraday's constant 96485.33, (A.s)/mol
I	electric current density, A
VF_w	volume fraction of pores filled by water in IEMs, –
D_i^s	ion diffusivity in the feed solutions in channels, m^2/s
r_i	radius of counter-ion i moving within the IEM, m
σ_m	membrane surface charge density, C/m^2
l_i^d	thickness of the double layer, m
ε	permittivity, F/m

Author details

Hesam Bazargan Harandi[1*] and Anahita Asadi[2]

1 State Key Laboratory of Geomechanics and Geotechnical Engineering, Institute of Rock and Soil Mechanics, Chinese Academy of Sciences, Wuhan, China

2 School of Automotive Engineering, Wuhan University of Technology, Wuhan, China

*Address all correspondence to: hbazargan@whut.edu.cn

References

[1] Harandi HB, Rahnama M, Jahanshahi Javaran E, Asadi A. Performance optimization of a multi stage flash desalination unit with thermal vapor compression using genetic algorithm. Applied Thermal Engineering. 2017;**123**: 1106-1119. DOI: 10.1016/j.applthermaleng.2017.05.170

[2] Harandi HB, Asadi A, Rahnama M, Shen Z-G, Sui P-C. Modeling and multi-objective optimization of integrated MED–TVC desalination system and gas power plant for waste heat harvesting. Computers and Chemical Engineering. 2021;**149**:107294. DOI: 10.1016/j.compchemeng.2021.107294

[3] El-Dessouky HT, Ettouney HM. Fundamentals of Salt Water Desalination. Amsterdam, The Netherlands: Elsevier; 2002

[4] Bazargan Harandi H, Asadi A, Shen Z, Sui P-C. Integration of direct-contact membrane distillation with flat-plate solar collector versus proton-exchange membrane fuel cell: Dynamic simulations and comparative analysis. Journal of Energy Engineering. 2022;**148**:1-15. DOI: 10.1061/(ASCE)EY.1943-7897.0000825

[5] El-Dessouky H, Alatiqi I, Bingulac S, Ettouney H. Steady-state analysis of the multiple effect evaporation desalination process. Chemical Engineering and Technology. 1998;**21**:437. DOI: 10.1002/(SICI)1521-4125(199805)21:5<437::AID-CEAT437>3.0.CO;2-D

[6] Hilal N, Khayet M, Wright CJ. Membrane Modification: Technology and Applications. Boca Raton, U.S.: CRC Press; 2012

[7] Misdan N, Lau WJ, Ismail AF. Seawater reverse osmosis (SWRO) desalination by thin-film composite membrane—Current development, challenges and future prospects. Desalination. 2012;**287**:228-237. DOI: 10.1016/j.desal.2011.11.001

[8] Sagle A, Freeman B. Fundamentals of membranes for water treatment. The Future of Desalination in Texas. 2004;**2**: 137. Available from: http://www.twdb.texas.gov/publications/reports/numbered_reports/doc/r363/c6.pdf

[9] Abdelrasoul A, Doan H, Lohi A, Cheng C-H. Mass transfer mechanisms and transport resistances in membrane separation process. InTech Open. 2015. DOI: 10.5772/60866

[10] Bazargan Harandi H, Asadi A, Shen Z, Rahnama M, Djilali N, Sui P-C. Modeling of heat and mass transfer in direct contact membrane distillation: Effect of counter diffusion velocity. Desalination and Water Treatment. 2021;**216**:71-82. DOI: 10.5004/dwt.2021.26816

[11] Zare S, Kargari A. Membrane properties in membrane distillation. In: Emerging Technologies for Sustainable Desalination Handbook. Oxford, United Kingdom: Elsevier; 2018. pp. 107-156

[12] Harandi HB, Zhang L, Kargari A, Sui P, Wang Y, Mutailipu M, et al. Optimum chain desalination process design for treatment of high TDS brine: A case assessment for future treatment of extracted brine from Shenhua CO_2 storage site. Greenhouse Gases: Science and Technology. 2023;**13**:470-492. DOI: 10.1002/ghg.2217

[13] Asadi A, Kang B, Harandi HB, Jung JC-Y, Shen Z, Sui P-C. Electrodialysis of lithium sulphate solution: Model development and validation. Journal of

the Electrochemical Society. 2022;**169**: 053508. DOI: 10.1149/1945-7111/ac707d

[14] Pathak N, Badeti U, Sohn W, Phuntsho S, Shon HK. Reverse osmosis (RO) membrane development and industrial applications. In: Current Developments in Biotechnology and Bioengineering. Amsterdam, The Netherlands: Elsevier; 2023. pp. 411-435. DOI: 10.1016/B978-0-443-19180-0.00021-3

[15] Walker JT, Surman-Lee S, McDermott PJ, Weinbren MJ. Reverse osmosis systems. In: Safe Water Healthc. United Kingdom: Elsevier; 2023. pp. 279-284. DOI: 10.1016/B978-0-323-90492-6.00029-X

[16] Sourirajan S. Reverse Osmosis. London, UK: Logos Press Ltd.; 1970

[17] Loeb S. The Loeb-Sourirajan Membrane: How it Came about. Washington, DC, U.S.: ACS Publications; 1981

[18] Yuster ST. Sea Water Demineralization by the "Surface Skimming" Process. California, U.S.: University of California, Department of Engineering; 1958

[19] Reid CE, Breton EJ. Water and ion flow across cellulosic membranes. Journal of Applied Polymer Science. 1959;**1**:133-143

[20] Loeb S, Sourirajan S. Sea Water Demineralization by Means of an Osmotic Membrane. Washington, DC, U.S.: ACS Publications; 1962

[21] Kaplan R, Mamrosh D, Salih HH, Dastgheib SA. Assessment of desalination technologies for treatment of a highly saline brine from a potential CO_2 storage site. Desalination. 2017;**404**: 87-101

[22] Ahmad M, Williams P. Assessment of desalination technologies for high saline brine applications—Discussion paper. Desalination and Water Treatment. 2011;**30**:22-36

[23] Bird R, Stewart W, Lightfoot E 2nd. Transport Phenomena. 2nd ed. Hoboken, NJ: John Wiley and Sons, Inc.; 2002

[24] Wijmans JG, Baker RW. The solution-diffusion model: A review. Journal of Membrane Science. 1995;**107**: 1-21

[25] Khayet M, Matsuura T. Membrane Distillation: Principles and Applications. Amsterdam: Elsevier; 2011

[26] Winter D. Membrane Distillation: A Thermodynamic, Technological and Economic Analysis. Freiburg im Breisgau, Germany: Shaker Verlag; 2015

[27] Harandi HB, Asadi A, Fathi H, Sui P-C. Combined macroscopic and pore scale modeling of direct contact membrane distillation with micro-porous hydrophobic membranes. Desalination. 2021;**514**:115171. DOI: 10.1016/j.desal.2021.115171

[28] Drioli E, Ali A, Macedonio F. Membrane distillation: Recent developments and perspectives. Desalination. 2015;**356**:56-84

[29] González D, Amigo J, Suárez F. Membrane distillation: Perspectives for sustainable and improved desalination. Renewable and Sustainable Energy Reviews. 2017;**80**:238-259

[30] Biniaz P, Torabi Ardekani N, Makarem M, Rahimpour M. Water and wastewater treatment systems by novel integrated membrane distillation (MD). ChemEngineering. 2019;**3**:8.

DOI: 10.3390/chemengineering 3010008

[31] Bodell BR. Silicone rubber vapor diffusion in saline water distillation (United States Pat. Ser). 1963

[32] Weyl PK. Recovery of Demineralized Water from Saline Waters. U.S.: A Google Patent; 1967

[33] Findley ME, Tanna VV, Rao YB, Yeh CL. Mass and heat transfer relations in evaporation through porous membranes. AICHE Journal. 1969;**15**:483-489

[34] Bodell BR. Distillation of Saline Water Using Silicone Rubber Membrane. U.S.: A Google Patent; 1968

[35] Cheng DY. Method and Apparatus for Distillation. U.S.: A Google Patent; 1984

[36] Esato K, Eiseman B. Experimental evaluation of Gore-Tex membrane oxygenator. The Journal of Thoracic and Cardiovascular Surgery. 1975;**69**:690-697

[37] Eleiwi F, Ghaffour N, Alsaadi AS, Francis L, Laleg-Kirati TM. Dynamic modeling and experimental validation for direct contact membrane distillation (DCMD) process. Desalination. 2016; **384**:1-11

[38] Adnan S, Hoang M, Wang H, Xie Z. Commercial PTFE membranes for membrane distillation application: Effect of microstructure and support material. Desalination. 2012;**284**:297-308

[39] Khayet M. Membranes and theoretical modeling of membrane distillation: A review. Advances in Colloid and Interface Science. 2011;**164**:56-88

[40] Qtaishat M, Rana D, Khayet M, Matsuura T. Preparation and characterization of novel hydrophobic/hydrophilic polyetherimide composite membranes for desalination by direct contact membrane distillation. Journal of Membrane Science. 2009;**327**:264-273

[41] Kang G, Cao Y. Application and modification of poly (vinylidene fluoride)(PVDF) membranes–a review. Journal of Membrane Science. 2014;**463**: 145-165

[42] Curcio E, Drioli E. Membrane distillation and related operations—A review. Separation & Purification Reviews. 2005;**34**:35-86

[43] Wang KY, Chung T-S, Gryta M. Hydrophobic PVDF hollow fiber membranes with narrow pore size distribution and ultra-thin skin for the fresh water production through membrane distillation. Chemical Engineering Science. 2008;**63**:2587-2594

[44] Zhang J, Dow N, Duke M, Ostarcevic E, Gray S. Identification of material and physical features of membrane distillation membranes for high performance desalination. Journal of Membrane Science. 2010;**349**:295-303

[45] Huang L-T, Hsu P-S, Kuo C-Y, Chen S-C, Lai J-Y. Pore size control of PTFE membranes by stretch operation with asymmetric heating system. Desalination. 2008;**233**:64-72

[46] Zhao C, Xue J, Ran F, Sun S. Modification of polyethersulfone membranes–a review of methods. Progress in Materials Science. 2013;**58**: 76-150

[47] Barzin J, Feng C, Khulbe KC, Matsuura T, Madaeni SS, Mirzadeh H. Characterization of polyethersulfone hemodialysis membrane by ultrafiltration and atomic force microscopy. Journal of Membrane Science. 2004;**237**:77-85

[48] Boussu K, Van der Bruggen B, Vandecasteele C. Evaluation of self-made nanoporous polyethersulfone membranes, relative to commercial nanofiltration membranes. Desalination. 2006;**200**:416-418

[49] Mulder M. Basic Principles of Membrane Technology. Dordrecht, The Netherlands: Kluwer Academic Publishers; 1996

[50] Roh IJ, Ramaswamy S, Krantz WB, Greenberg AR. Poly (ethylene chlorotrifluoroethylene) membrane formation via thermally induced phase separation (TIPS). Journal of Membrane Science. 2010;**362**:211-220

[51] Dickey CA, Mcdaniel JE. Method of Producing Spherical Thermoplastic Particles. U.S.: A Google Patent; 1975

[52] Sadeghi F. Development of Microporous Polypropylene by Stretching, Ecole Polytechnique de Montreal. Ottawa, ON, Canada: PolyPublie Polytechnique Montreal; 2006

[53] Sarada T, Sawyer LC, Ostler MI. Three dimensional structure of celgard® microporous membranes. Journal of Membrane Science. 1983;**15**:97-113

[54] Zhu W, Zhang X, Zhao C, Wu W, Hou J, Xu M. A novel polypropylene microporous film. Polymers for Advanced Technologies. 1996;**7**:743-748

[55] Ulbricht M. Advanced functional polymer membranes. Polymer (Guildf). 2006;**47**:2217-2262

[56] Drioli E, Giorno L. Membrane Operations: Innovative Separations and Transformations. Weinheim, Germany: Wiley-VCH; 2009

[57] Hu J, Harandi HB, Chen Y, Zhang L, Yin H, He T. Anisotropic gypsum scaling of corrugated polyvinylidene fluoride hydrophobic membrane in direct contact membrane distillation. Water Research. 2023;**244**:120513. DOI: 10.1016/j.watres.2023.120513

[58] Prince JA, Singh G, Rana D, Matsuura T, Anbharasi V, Shanmugasundaram TS. Preparation and characterization of highly hydrophobic poly (vinylidene fluoride)–clay nanocomposite nanofiber membranes (PVDF–clay NNMs) for desalination using direct contact membrane distillation. Journal of Membrane Science. 2012;**397**:80-86

[59] Ray SS, Chen S-S, Li C-W, Nguyen NC, Nguyen HT. A comprehensive review: Electrospinning technique for fabrication and surface modification of membranes for water treatment application. RSC Advances. 2016;**6**:85495-85514

[60] Bazargan Harandi H, Hu J, Asadi A, Sui P-C, Zhang L, He T. Experimental and theoretical analysis of scaling mitigation for corrugated PVDF membranes in direct contact membrane distillation. Journal of Membrane Science. 2023;**686**:122001. DOI: 10.1016/j.memsci.2023.122001

[61] Bernardes A, Rodrigues MAS, Ferreira JZ. Electrodialysis and Water Reuse. Heidelberg, New York, Dordrecht, London: Springer; 2016

[62] Strathmann H. Ion-Exchange Membrane Separation Processes. Amsterdam: Elsevier; 2004

[63] Paidar M, Fateev V, Bouzek K. Membrane electrolysis—History, current status and perspective. Electrochimica Acta. 2016;**209**:737-756

[64] Al-Amshawee S, Yunus MYBM, Azoddein AAM, Hassell DG, Dakhil IH, Hasan HA. Electrodialysis desalination for water and wastewater: A review. Chemical Engineering Journal. 2020;**380**: 122231

[65] Campione A, Gurreri L, Ciofalo M, Micale G, Tamburini A, Cipollina A. Electrodialysis for water desalination: A critical assessment of recent developments on process fundamentals, models and applications. Desalination. 2018;**434**:121-160

[66] Ali A, Tufa RA, Macedonio F, Curcio E, Drioli E. Membrane technology in renewable-energy-driven desalination. Renewable and Sustainable Energy Reviews. 2018;**81**:1-21

[67] Grebenyuk VD, Grebenyuk OV. Electrodialysis: From an idea to realization. Russian Journal of Electrochemistry. 2002;**38**:806-809

[68] Wyllie MRJ, Patnode HW. The development of membranes prepared from artificial cation-exchange materials with particular reference to the determination of sodium-ion activities. The Journal of Physical Chemistry. 1950; **54**:204-227

[69] Jiang C, Hossain MM, Li Y, Wang Y, Xu T. Ion exchange membranes for Electrodialysis: A comprehensive review of recent advances. Journal of Membranes and Separation Technology. 2014;**3**:185

[70] Sollner K, Dray S, Grim E, Neihof R. Electrochemical studies with model membranes. In: Ion Transport Across Membranes. New York: Academic Press; 1954. pp. 144-188

[71] Frilette VJ. Preparation and characterization of bipolar ion exchange membranes. The Journal of Physical Chemistry. 1956;**60**:435-439

[72] Reahl ER. Half a Century of Desalination with Electrodialysis. U.S.: General Electric Company; 2004

[73] Hotel WP. The 41st International Water Conference. Pittsburgh, Pennsylvania, USA: Engineers' Society of Western Pennsylvania; 1980

[74] Dursun L, Ata ON, Kanca A. Bipolar membrane electrodialysis for binary salt water treatment: Valorization of type and concentration of electrolytes. Industrial and Engineering Chemistry Research. 2021;**60**:2003-2010

[75] Fernandez-Gonzalez C, Dominguez-Ramos A, Ibañez R, Irabien A. Electrodialysis with bipolar membranes for valorization of brines. Separation and Purification Reviews. 2016;**45**: 275-287

[76] Jaroszek H, Dydo P. Ion-exchange membranes in chemical synthesis–a review. Open Chemistry. 2016;**14**: 1-19

[77] Valero F, Barceló A, Arbós R. Electrodialysis technology: Theory and applications. Desalination, Trends and Technologies. 2011;**28**:3-20

[78] Strathmann H. Electrodialysis and its application in the chemical process industry. Separation and Purification Methods. 1985;**14**:41-66

[79] Asadi A, Bazargan Harandi H, Kang B, Jung JC-Y, Sui P-C. A comprehensive computational fluid dynamics modeling of lithium sulphate electrodialysis. Journal of the Electrochemical Society. 2023;**170**:1-13. DOI: 10.1149/1945-7111/acf529

[80] Tanaka Y. Ion Exchange Membranes: Fundamentals and Applications. Amsterdam: Elsevier; 2015

[81] Strathmann H. Ion-exchange membrane processes in water treatment. Sustainability Science and Engineering. 2010;**2**:141-199

[82] Sata T. Ion Exchange Membranes: Preparation, Characterization, Modification and Application. United Kingdom: Royal Society of Chemistry; 2007

[83] Honarparvar S, Reible D. Modeling multicomponent ion transport to investigate selective ion removal in electrodialysis. Environmental Science and Ecotechnology. 2020;**1**:100007

[84] Jiang C, Wang Q, Li Y, Wang Y, Xu T. Water electro-transport with hydrated cations in electrodialysis. Desalination. 2015;**365**:204-212

[85] Donnan FG, Guggenheim EA. Die genaue thermodynamik der membrangleichgewichte. Zeitschrift für Physikalische Chemie. 1932;**162**:346-360

[86] Zourmand Z, Faridirad F, Kasiri N, Mohammadi T. Mass transfer modeling of desalination through an electrodialysis cell. Desalination. 2015;**359**:41-51

[87] Generous MM, Qasem NAA, Zubair SM. The significance of modeling electrodialysis desalination using multi-component saline water. Desalination. 2020;**496**:114347

[88] Chein R, Chen H, Liao C. Investigation of ion concentration and electric potential distributions in charged membrane/electrolyte systems. Journal of Membrane Science. 2009;**342**:121-130

[89] Taky M, Pourcelly G, Lebon F, Gavach C. Polarization phenomena at the interfaces between an electrolyte solution and an ion exchange membrane: Part I. Ion transfer with a cation exchange membrane. Journal of Electroanalytical Chemistry. 1992;**336**:171-194

[90] Wisniewska G, Winnicki T. Electrodialytic desalination of effluents from zinc-coating processes: Removal of Zn^{2+} and CL^{-} ions from model solutions. Desalination. 1991;**84**:163-176